REVEALING THE HIDDEN NATURE OF
SPACE AND TIME

Charting the Course for Elementary Particle Physics

Committee on Elementary Particle Physics in the 21st Century

Board on Physics and Astronomy

Division on Engineering and Physical Sciences

NATIONAL RESEARCH COUNCIL
OF THE NATIONAL ACADEMIES

THE NATIONAL ACADEMIES PRESS
Washington, D.C.
www.nap.edu

THE NATIONAL ACADEMIES PRESS • 500 Fifth Street, N.W. • Washington, DC 20001

NOTICE: The project that is the subject of this report was approved by the Governing Board of the National Research Council, whose members are drawn from the councils of the National Academy of Sciences, the National Academy of Engineering, and the Institute of Medicine. The members of the committee responsible for the report were chosen for their special competences and with regard for appropriate balance.

This study was supported by Grant No. PHY-0432486 between the National Academy of Sciences and the National Science Foundation and Contract No. DE-FG02-04ER41327 between the National Academy of Sciences and the Department of Energy. Any opinions, findings, conclusions, or recommendations expressed in this publication are those of the author(s) and do not necessarily reflect the views of the organizations or agencies that provided support for the project.

Library of Congress Cataloging-in-Publication Data

National Research Council (U.S.). Committee on Elementary Particle Physics in the 21st Century.
  Revealing the hidden nature of space and time : charting the course for elementary particle physics / Committee on Elementary Particle Physics in the 21st Century, Board on Physics and Astronomy, Division on Engineering and Physical Sciences.
      p. cm.
  Includes bibliographical references.
   ISBN 0-309-10194-8 (pbk.) — ISBN 0-309-66039-4 (pdf) 1. Particles (Nuclear physics)—Research—United States. 2. Space and time—Research—United States. I. Title.
QC793.4.N38 2006
539.7'2072073—dc22

                        2006027444

Additional copies of this report are available from the National Academies Press, 500 Fifth Street, N.W., Lockbox 285, Washington, DC 20055; (800) 624-6242 or (202) 334-3313 (in the Washington metropolitan area); Internet, http://www.nap.edu.

THE NATIONAL ACADEMIES
Advisers to the Nation on Science, Engineering, and Medicine

The **National Academy of Sciences** is a private, nonprofit, self-perpetuating society of distinguished scholars engaged in scientific and engineering research, dedicated to the furtherance of science and technology and to their use for the general welfare. Upon the authority of the charter granted to it by the Congress in 1863, the Academy has a mandate that requires it to advise the federal government on scientific and technical matters. Dr. Ralph J. Cicerone is president of the National Academy of Sciences.

The **National Academy of Engineering** was established in 1964, under the charter of the National Academy of Sciences, as a parallel organization of outstanding engineers. It is autonomous in its administration and in the selection of its members, sharing with the National Academy of Sciences the responsibility for advising the federal government. The National Academy of Engineering also sponsors engineering programs aimed at meeting national needs, encourages education and research, and recognizes the superior achievements of engineers. Dr. Wm. A. Wulf is president of the National Academy of Engineering.

The **Institute of Medicine** was established in 1970 by the National Academy of Sciences to secure the services of eminent members of appropriate professions in the examination of policy matters pertaining to the health of the public. The Institute acts under the responsibility given to the National Academy of Sciences by its congressional charter to be an adviser to the federal government and, upon its own initiative, to identify issues of medical care, research, and education. Dr. Harvey V. Fineberg is president of the Institute of Medicine.

The **National Research Council** was organized by the National Academy of Sciences in 1916 to associate the broad community of science and technology with the Academy's purposes of furthering knowledge and advising the federal government. Functioning in accordance with general policies determined by the Academy, the Council has become the principal operating agency of both the National Academy of Sciences and the National Academy of Engineering in providing services to the government, the public, and the scientific and engineering communities. The Council is administered jointly by both Academies and the Institute of Medicine. Dr. Ralph J. Cicerone and Dr. Wm. A. Wulf are chair and vice chair, respectively, of the National Research Council.

www.national-academies.org

Preface

The principal charge to the Committee on Elementary Particle Physics in the 21st Century was to recommend priorities for the U.S. particle physics program for the next 15 years. Described in the Executive Summary and more fully presented in the Overview, the committee's considered response is laid out in detail in the main text of this report, which begins by discussing the scientific challenges in particle physics and conveying the current status of the U.S. program and then presents the committee's consensus on the best way to sustain a competitive and globally relevant U.S. particle physics program.

Given the charge (see Appendix B), the composition of this committee was something of an experiment for the National Academies. The committee membership went well beyond particle physicists and accelerator scientists to include condensed matter physicists, astrophysicists, astronomers, biologists, industrialists, and a variety of experts in public policy, particularly science policy. As a result, a good deal of education was necessary during the course of the study, and we members who are not particle physicists would like to express our gratitude for the intellectual generosity and patience of the committee's physicists as they provided us with the level of understanding necessary to proceed with the task. In the same vein, for their considerable assistance the committee owes a great deal to its colleagues at the major particle physics laboratories in the United States (Argonne National Laboratory, Brookhaven National Laboratory, Cornell Laboratory for Elementary Particle Physics, Fermi National Accelerator Laboratory, Lawrence

Berkeley National Laboratory, and Stanford Linear Accelerator Center) and to colleagues abroad at the Deutsches Elektronen-Synchroton (DESY) laboratory in Hamburg, the European Center for Nuclear Research (CERN) laboratory in Geneva, and the Japan Proton Accelerator Research Complex (J-PARC) and High Energy Accelerator Research Organization (KEK) laboratories in Japan.

For the nonphysicists on the committee, the task was both intellectually exciting and sobering. Simply stated, we nonphysicists were not fully aware of the challenge faced by the U.S. particle physics program in sustaining its tradition of leadership. Given the globalization of particle physics (and with Europe investing twice as much as the United States and Japan investing nearly half as much as the United States in particle physics), identifying a compelling leadership role for the United States was not simple. Since the unfortunate demise of the Superconducting Super Collider in the early 1990s and the subsequent stagnation of support for U.S. efforts in particle physics, the U.S. program has lacked a long-term and distinguishing strategic focus that would give it a competitive and distinctive position within the worldwide effort in particle physics. The entire committee came to believe that it was essential to adopt a compelling set of national priorities within a well-defined, long-term strategic framework. Equally important, the committee accepted the need for the United States to shoulder some risk in order to maximize the opportunity to meet the leadership and scientific challenges in particle physics.

With respect to the unusual composition of the committee (see Appendix D), others will judge whether this experiment should be repeated, but it is our judgment that all members of the committee contributed distinctive and important perspectives that helped the group as a whole to devise a more compelling set of recommendations. In particular, members from outside particle physics posed challenging questions to those inside the field and listened carefully to the arguments. The result was an overall sharpening of everyone's thinking as well as stronger connections to a broader context.

Finally, we both want to personally acknowledge and thank every committee member for the tremendous attention and effort each devoted to this activity. Some members traveled great distances to participate in the committee meetings, and everyone made difficult choices about other commitments to make this project a key priority. It is only through these generous and combined efforts that this report achieved clarity and closure.

Harold T. Shapiro, *Chair*
Sally Dawson, *Vice Chair*
Committee on Elementary Particle Physics
in the 21st Century

Acknowledgments

T his report is the product of large amounts of work by many people. The committee extends its thanks and appreciation to all who participated in this endeavor; it would be impossible to name them all individually.

The committee thanks the speakers who made formal presentations at each of the meetings; their presentations and the ensuing discussions were extremely informative and had a significant impact on the committee's deliberations. The committee is especially appreciative of efforts by members of the international community (Robert Aymar, Ian Halliday, Yoji Totsuka, and Albrecht Wagner) to participate in its May 2005 meeting in Illinois and its August 2005 meeting in New York. And in general, the committee acknowledges the extra work required to prepare remarks addressing the broad spectrum of expertise on the committee.

The committee also expresses its deep gratitude to the hosts and facilitators for each of its meetings at the particle physics laboratories in the United States (Jonathan Dorfan at the Stanford Linear Accelerator Center (SLAC), Michael Witherell and Piermaria Oddone at Fermilab, and Maury Tigner at Cornell). Most especially, the committee is grateful for the hospitality and warmth of its hosts at site visits abroad (Robert Aymar at CERN, Brian Foster of the United Kingdom, Shoji Nagamiya at J-PARC, Roberto Petronzio of the Istituto Nazionale di Fisica Nucleare (INFN), Yoji Totsuka at KEK, and Albrecht Wagner at DESY). These visits overseas were tremendously valuable.

The committee also thanks those who sent in letters and e-mail messages in response to questions posed by the committee. In particular, the excellent efforts of the Lykken/Siegrist subpanel of the High Energy Physics Advisory Panel were very helpful.

Finally, the committee thanks the staff of the Board on Physics and Astronomy (Donald Shapero, Timothy Meyer, and David Lang) for their guidance and assistance throughout this process.

Acknowledgment of Reviewers

This report has been reviewed in draft form by individuals chosen for their diverse perspectives and technical expertise, in accordance with procedures approved by the National Research Council's Report Review Committee. The purpose of this independent review is to provide candid and critical comments that will assist the institution in making its published report as sound as possible and to ensure that the report meets institutional standards for objectivity, evidence, and responsiveness to the study charge. The review comments and draft manuscript remain confidential to protect the integrity of the deliberative process. We wish to thank the following individuals for their review of this report:

W.F. Brinkman, Princeton University
Persis Drell, Stanford Linear Accelerator Center
Ralph Eichler, Paul Scherrer Institute
Paul H. Gilbert, Parsons Brinckerhoff, Inc.
Ian Halliday, European Science Foundation and Scottish Universities
    Physics Alliance, University of Edinburgh
Wick C. Haxton, University of Washington
Bernadine P. Healy, U.S. News and World Report
Rolf-Dieter Heuer, Deutsches Elektronen-Synchrotron, DESY
John P. Huchra, Harvard-Smithsonian Center for Astrophysics
Christopher Llewellyn-Smith, United Kingdom Atomic Energy Authority,
    Culham Division

Joseph Lykken, Fermi National Accelerator Laboratory
Satoshi Ozaki, Brookhaven National Laboratory
John Peoples, Fermi National Accelerator Laboratory
Burton Richter, Stanford Linear Accelerator Center
Yoji Totsuka, High Energy Accelerator Research Organization, KEK
Charles M. Vest, Massachusetts Institute of Technology
Bruce D. Winstein, University of Chicago

Although the reviewers listed above have provided many constructive comments and suggestions, they were not asked to endorse the conclusions or recommendations, nor did they see the final draft of the report before its release. The review of this report was overseen by Louis J. Lanzerotti of the New Jersey Institute of Technology and William Happer of Princeton University. Appointed by the National Research Council, they were responsible for making certain that an independent examination of this report was carried out in accordance with institutional procedures and that all review comments were carefully considered. Responsibility for the final content of this report rests entirely with the authoring committee and the institution.

Contents

EXECUTIVE SUMMARY   1

OVERVIEW   5

1   THE SCIENTIFIC EXCITEMENT AND CHALLENGES   17
    Challenges to the Standard Model, 20
    Responding to the Challenges, 24
    The Role of the United States in Particle Physics, 26

2   KEY QUESTIONS IN PARTICLE PHYSICS   33
    Can All the Forces Between Particles Be Understood in a Unified
        Framework?, 34
    What Do the Properties of Particles Reveal About the Nature and
        Origin of Matter and the Properties of Space and Time?, 39
    What Are Dark Energy and Dark Matter and How Has Quantum
        Mechanics Influenced the Structure of the Universe?, 50
    Roles of Accelerator- and Non-Accelerator-Based Experiments, 53

3   THE EXPERIMENTAL OPPORTUNITIES   56
    High-Energy Beams: Direct Exploration of the Terascale, 57
        Discoveries at the Terascale, 57
        Tools for Exploring the Terascale, 57

Physics at the Terascale, 64
Toward the Terascale, 75
High-Intensity Beams, 77
Nature's Particle Sources, 84
International Cooperation, 89
    Global Activity in Particle Physics, 89
    The International Linear Collider, 93
A Path Forward, 94
Opportunities Ahead, 97

4   THE STRATEGIC FRAMEWORK   101
    The Scientific Challenge, 101
    The Position of the U.S. Program, 102
    The Strategic Principles, 104
    The Budgetary Framework, 111
        Recent Trends in Support for the U.S. Particle Physics Program, 111
        Multiyear Plans and Budgets, 112
    National Program Considerations, 113
    Budget Considerations, 115

5   FINDINGS AND RECOMMENDED ACTIONS   118
    The Scientific Agenda for Elementary Particle Physics, 118
    Priorities, 119
    Direct Exploration of the Terascale, 119
    Explorations of Particle Astrophysics and Unification, 129
    Implications of the Strategic Agenda Under Different Budget Scenarios, 133
    Realizing the Strategic Vision for Elementary Particle Physics, 135

AFTERWORD   136

APPENDIXES

A   International Progress Toward the ILC   139
B   Charge to the Committee   143
C   Committee Meeting Agendas   144
D   Biographical Sketches of Committee Members and Staff   152

Executive Summary

A national discussion about the future of U.S. global leadership in science, technology, and innovation has been unfolding over the past few years. In October 2005, echoing widespread concerns,[1] the report *Rising Above the Gathering Storm* outlined a program designed to enhance the U.S. science and technology enterprise so that the nation can sustain its cultural vitality, continue to provide leadership, and successfully compete, prosper, and be secure in an increasingly globalized world. In particular, the report identified basic research in the physical sciences as a key underpinning for the nation's strategic strengths.

Against this broader backdrop, the work of the Committee on Elementary Particle Physics in the 21st Century took on a special significance. By recognizing the need for U.S. leadership in particle physics, and by articulating an approach to ensuring that leadership, this report offers a compelling opportunity for action in the national discussion of the U.S. role in science and technology. Simply stated,

---

[1]See, for example, House Committee on Science, *Unlocking Our Future: Toward a New National Science Policy*, September 1998, available online at <http://www.house.gov/science/science_policy_report.htm>; T.L. Friedman, *The World Is Flat: A Brief History of the Twenty-first Century*, New York: Farrar, Strauss, and Giroux, 2005; National Academy of Sciences (NAS), National Academy of Engineering (NAE), and Institute of Medicine (IOM), *Rising Above the Gathering Storm: Energizing and Employing America for a Brighter Economic Future*, Washington, D.C.: The National Academies Press, 2005 (Prepublication); U.S. Domestic Policy Council, *American Competitiveness Initiative*, February 2006.

given the excitement of the scientific opportunities in particle physics, and in keeping with the nation's broader commitment to research in the physical sciences, the committee believes that the United States should continue to support a competitive program in this key scientific field.

However, despite the sense of excitement and anticipation within particle physics, the U.S. tradition of leadership in the field is not secure. The major U.S. particle physics experimental facilities are entering an era of change, with some facilities being closed and others transitioning to new purposes, and support for particle physics in the United States has stagnated. As a result, the intellectual center of gravity within the field is moving abroad. Within a few years, a majority of U.S. experimental particle physicists will be involved in experiments being conducted in other countries.

The U.S. program in particle physics is at a crossroads. The continuing vitality of the program requires new, decisive, and forward-looking actions. In addition, sustained leadership requires a willingness to take the risks that always accompany leadership on the scientific frontier. Thus, the committee recommends the thoughtful pursuit of a high-risk, high-reward strategy.

The most important components of such a strategy are the establishment of a set of important new experiments in the United States (including a large accelerator facility), a determination to work together with colleagues abroad in mutually beneficial joint ventures, adoption of a compelling set of priorities within a broad strategic framework, and the provision of reasonable levels of resources. The committee particularly emphasizes the increasing benefits of establishing cooperative ventures with programs in other countries, whether the experimental facilities are located in the United States or abroad. These joint ventures will provide U.S. students and scientists with a full range of exciting scientific opportunities and meet the obligation to deploy public funds responsibly.

The committee arrived at three strong conclusions regarding both particle physics and the U.S. role in this global scientific and technological enterprise:

1. Particle physics plays an essential role in the broader enterprise of the physical sciences. It inspires U.S. students, attracts talent from around the world, and drives critical intellectual and technological advances in other fields.

2. Although setting priorities is essential, it also is critical to maintain a diverse portfolio of activities in particle physics, from theory to accelerator R&D to the construction and support of new experimental facilities. The committee believes that accelerators will remain an essential component of the program, since some critical scientific questions cannot be explored in any other manner.

3. The field of elementary particle physics is entering an era of unprecedented potential. New experimental facilities, including accelerators, space-based experiments, underground laboratories, and critical precision measurements of various kinds, offer a variety of ways to explore the hidden nature of matter, energy, space, and time. The availability of technologies that can explore directly an energy regime known as the Terascale is especially exciting. The direct exploration of the Terascale could be the next important step toward resolving questions that human beings have asked for millennia: What are the origins of mass? Can the basic forces of nature be unified? How did the universe begin? How will it evolve in the future? Moreover, at Terascale energies, formerly separate questions in cosmology and particle physics become connected, bridging the sciences of the very large and the very small.

The results of the committee's analysis have led to its chief recommendation:

The United States should remain globally competitive in elementary particle physics by playing a leading role in the worldwide effort to aggressively study Terascale physics.

To implement the committee's chief recommendation, the Department of Energy and the National Science Foundation should work together to achieve the following objectives in priority order:

1. Fully exploit the opportunities afforded by the construction of the Large Hadron Collider (LHC) at the European Center for Nuclear Research (CERN).
2. Plan and initiate a comprehensive program to become the world-leading center for research and development on the science and technology of a linear collider, and do what is necessary to mount a compelling bid to build the proposed International Linear Collider (ILC) on U.S. soil.
3. Expand the program in particle astrophysics and pursue an internationally coordinated, staged program in neutrino physics.

The LHC will begin exploratory research at the Terascale within the next few years. Physicists expect it to produce evidence for the Higgs particle that is hypothesized to be responsible for generating the mass of all matter. In addition, theoretical arguments point to the possibility of discovering a new symmetry, known as supersymmetry, at the LHC in the form of new particles that are partners to the currently known particles; some of these new supersymmetric particles may turn out to constitute the mysterious "dark matter" that pervades the universe.

When the LHC has outlined the territory of Terascale physics, more precise and sensitive measurements will be needed. For that purpose, a new accelerator facility that collides electrons and positrons will be required. The committee believes that the United States should invest the capital needed to host the proposed ILC as the essential component of U.S. leadership in particle physics in the decades ahead.

The committee recognizes that more than one strategy could be pursued in the next decade, but in its judgment the priorities it has outlined have the highest risk-adjusted return and constitute the strategy most likely to sustain U.S. leadership in particle physics.

The next few decades will represent a culmination of the human effort to understand the elementary constituents of the universe. The United States has an unprecedented opportunity, as a leader of nations, to undertake this profound scientific challenge.

Overview

THE SCIENCE OPPORTUNITIES

Elementary particle physics—the study of the fundamental constituents and nature of the universe—is poised to take the next significant step in answering questions that humans have asked for millennia: What is the nature of space and time? What are the origins of mass? How did the universe begin? How will it evolve in the future? The next few decades could be one of the most exciting periods in the history of physics.

One of the great scientific achievements of the 20th century was the development of the Standard Model of elementary particle physics, which describes the relationships among the known elementary particles and the characteristics of three of the four forces that act on those particles—electromagnetism, the strong force, and the weak force (but not gravity). However, in the energy regions that physicists are just now becoming able to access experimentally, the incompleteness of the Standard Model becomes apparent. It is unable to reconcile the twin pillars of 20th century physics, Einstein's general theory of relativity and quantum mechanics. In addition, recent astronomical observations indicate that everyday matter accounts for just 4 percent of the total substance in the universe. The rest of the universe consists of hypothesized entities called dark matter and dark energy that are not described by the Standard Model. Other challenges to the Standard Model are posed by the predominance of matter over antimatter in the universe, the early evolution of the universe, and the discovery that the elusive particles known as neutrinos have a tiny but nonzero mass. Thus, despite the extraordinary success of the Standard Model, it seems likely that a much deeper understanding of nature will be achieved as physicists continue to study the fundamental constituents of the universe.

Elementary particle physicists use a wide variety of natural phenomena to investigate the properties and interactions of particles. They gather data from cosmic rays and solar neutrinos, astronomical observations, precision measurements of single particles, and monitoring of large masses of everyday matter. In addition, crucial advances historically have come from particle accelerators and the complex detectors used to study particle collisions in controlled environments. Today the most powerful accelerator in the world is the Tevatron at the Fermi National Accelerator Laboratory (Fermilab) in Batavia, Illinois, which is scheduled to be shut down by the end of the decade. A more powerful accelerator, the Large Hadron Collider (LHC) at the European Center for Nuclear Research (CERN) in Geneva, Switzerland, is scheduled to begin colliding protons in 2007. Both theoretical and experimental evidence suggests that revolutionary new physics will emerge at the energies accessible with the LHC.

Beyond the LHC, physicists around the world are designing a new accelerator known as the International Linear Collider (ILC), which would use two linear accelerators to collide beams of electrons and positrons. Together, the LHC and an ILC will enable physicists to explore the unification of the fundamental forces, probe the origins of mass, uncover the dynamic nature of the "vacuum" of space, deepen the understanding of stellar and nuclear processes, and investigate the nature of dark matter. These tasks cannot be accomplished with the LHC alone.

THE U.S. ROLE IN PARTICLE PHYSICS

For more than half a century, the United States has been a leader in particle physics. But over the next few years, as the flagship U.S. particle physics facilities are surpassed on the energy frontier by new facilities overseas or are converted to other uses, the intellectual center of gravity of the field will move abroad. At the same time, the conclusion of these important experiments creates an opportunity for the United States to consider major new initiatives.

Today, the U.S. program in elementary particle physics is at a crossroads. For the U.S. program to remain relevant in the global context, it must take advantage of exciting new opportunities. Doing so will require decisive actions and strong commitments; it also will require a willingness to assume some risks. Thus, to ensure continued U.S. leadership in this important scientific area, a new strategic framework is needed that can guide the difficult decisions that have to be made.

STRATEGIC PRINCIPLES

Seven strategic principles underlie the actions recommended by the committee:

Strategic Principle 1. The committee affirms the intrinsic value of elementary particle physics as part of the broader scientific and technological enterprise and identifies it as a key priority within the physical sciences.

A strong role in particle physics is necessary if the United States is to sustain its leadership in science and technology over the long term. The nation's investments in basic research in the physical sciences have contributed greatly to U.S. scientific and technological prowess. Elementary particle physics has been a centerpiece of the physical sciences throughout the 20th century. It has inspired generations of young people to become members of the strongest scientific workforce in the world. It also has attracted outstanding scientists from abroad to come to the United States and contribute to the nation's intellectual and economic vitality.

In addition, particle physics has generated waves of technological innovations that have found applications throughout the sciences and society. The protocols that underlie the World Wide Web were developed at CERN, and the two-way interactions between particle physics and high-performance computing and communications have continued to blossom. Particle physics has generated critical technologies in such areas as materials analysis, medical treatment, and imaging.

Strategic Principle 2. The U.S. program in elementary particle physics should be characterized by a commitment to leadership within the global particle physics enterprise.

In today's world, leadership in the sciences does not mean singular dominance. Rather, leadership is characterized by taking initiatives on the scientific frontier, accepting risks, and catalyzing partnerships with colleagues at home and abroad. A leadership position enables a country to exploit scientific and technological developments no matter where they emerge. The U.S. program should not only pursue the most compelling scientific opportunities, but it also should establish a clear path for the United States to reach a position of leadership in particle physics.

Strategic Principle 3. As the global particle physics research program becomes increasingly integrated, the U.S. program in particle physics should be planned and executed with greater emphasis on strategic international partnerships. The United States should lead in mobilizing the interests of international partners to jointly plan, site, and sponsor the most effective and the most important experimental facilities.

As experimental facilities become more complex and expensive, the already extensive levels of international collaboration in particle physics will need to intensify further to most effectively address the challenges on the scientific frontier. The committee believes that particle physics should evolve into a truly global collaboration that would enable the particle physics community to leverage its resources, prevent duplication of effort, and maximize opportunities for particle physicists throughout the world. Credible and reliable participation, as well as leadership, in strategic international partnerships require the United States to maintain a healthy and vital particle physics program.

Strategic Principle 4. The committee believes that the U.S. program in elementary particle physics must be characterized by the following to achieve and sustain a leadership position. Together, these characteristics provide for a program in particle physics that will be lasting and continuously beneficial:

- **A long-term vision,**
- **A clear set of priorities,**
- **A willingness to take scientific risks where justified by the potential for major advances,**
- **A determination to seek mutually advantageous joint ventures with colleagues abroad,**
- **A considerable degree of flexibility and resiliency,**
- **A budget consistent with an aspiration for leadership, and**
- **As robust and diversified a portfolio of research efforts as investment levels permit.**

The last of these characteristics—breadth—deserves special consideration. A broad array of scientific opportunities exists in elementary particle physics, and it is not possible to foretell which will yield important new results soonest. Two of the greatest discoveries of the last decade—those of nonzero neutrino masses and dark energy—were quite unexpected and arose from experiments that did not use accelerators, the tools characteristic of many other advances in particle physics. Thus, there is a strong need for supporting a variety of approaches to current scientific opportunities.

It is important to maintain a diverse and comprehensive portfolio of research activities that encompasses university-based students and faculty, national laboratories, and activities conducted in other countries. Even during periods of budgetary stringency, sufficient funding and diversity must be retained in the pipeline of projects so that the United States is positioned to participate in the most exciting science wherever it occurs.

Strategic Principle 5. The Secretary of Energy and the Director of the National Science Foundation, working with the White House Office of Science and Technology Policy and the Office of Management and Budget and in consultation with the relevant authorization and appropriations committees of Congress, should, as a matter of strategic policy, establish a 10- to 15-year budget plan for the elementary particle physics program.

Many important experiments in particle physics require multiyear plans and budgets. Experience with past science projects has shown that uncertainties and shortfalls in annual appropriations can lead to unnecessary cost escalations and to inefficient and unwise, even if expeditious, decisions. The ability to make sustained multiyear commitments is also essential if the United States is to appear credible and serious in the international arena, especially in terms of fostering collaboration and cooperation.

Strategic Principle 6. A strong and vital Fermilab is an essential element of U.S. leadership in elementary particle physics. Fermilab must play a major role in advancing the priorities identified in this report.

Many universities and national laboratories have made vital contributions to particle physics over the years. But in recent years the number of laboratories devoted primarily to particle physics has been declining and will continue to do so, especially as the facilities at the Stanford Linear Accelerator Center and at Cornell University direct their primary focus away from particle physics. Continuing efforts from university groups and other laboratories will be essential to realize the full potential of the U.S. particle physics program. At the same time, Fermilab will play a special role as the only laboratory dedicated chiefly to particle physics.

Strategic Principle 7. A standing national program committee should be established to evaluate the merits of specific projects and to make recommendations to DOE and NSF regarding the national particle physics program in the context of international efforts.

The changing environment in particle physics requires a reexamination of the advisory structure for the field. The combination of unparalleled opportunities in particle physics and inevitable fiscal constraints force the federal government and the particle physics community to make very hard choices and coordinate programs at the various national laboratories and universities. A standing national committee is needed that has sufficient authority to establish a compelling set of priorities and to advise the federal agencies that support particle physics. Such a committee should evaluate the merits of specific proposals and make recommendations regarding the national particle physics program within the context of the international particle physics program. Existing advisory committees such as the Department of Energy (DOE)/National Science Foundation (NSF) High Energy Physics Advisory Panel (HEPAP) or the Particle Physics Project Prioritization Panel (P5) could be strengthened and broadened to take on this role.

RECOMMENDED ACTION ITEMS

The committee examined several possible scenarios for the funding of particle physics in the United States. Much of the analysis for the next few years was conducted assuming a budget that would rise with the rate of inflation, representing a constant level of effort (though particle physics would represent an ever smaller proportion of the gross domestic product). If, instead, the budget remains flat and without any adjustments for inflation, policy makers will have decided to disinvest in this area of science. This course is incompatible with the goal of leadership for the U.S. program in particle physics.

Recently, both the executive and the legislative branches of the federal government expressed a desire to increase funding for basic research in the physical sciences. Real increases ranging from 2 to 3 percent per year to a doubling over 7 years would enable many exciting experiments to be conducted that cannot be realized in the constant-effort budget.

The committee presents its recommended strategy for the U.S. role in particle physics over the next 15 years in the form of six action items ranked in priority order. The most compelling current scientific opportunity in elementary particle physics is exploration of the Terascale, and this is the committee's highest priority for the U.S. program. Direct investigations of phenomena at the energy frontier hold the greatest promise for transformational advances. Within this context, the experimental programs at the LHC and at the proposed ILC offer the best means for seizing this opportunity.

The committee's recommended strategy for exploitation of the LHC and initiation of the ILC addresses projects at radically different stages of realization. On the one hand, the construction phase of the LHC project, including the installation of its massive detectors, is essentially complete, and the global particle physics community is ready to use it. On the other hand, the ILC remains a concept in development, although a substantial amount of R&D demonstrating the feasibility of the technologies selected for the facility has been successfully undertaken during the past decade. Taken together, these two facilities represent a 20-year campaign to seize the opportunities afforded by the opening of the Terascale.

Action Item 1. The highest priority for the U.S. national effort in elementary particle physics should be to continue as an active partner in realizing the physics potential of the LHC experimental program.

The LHC will be the center of gravity for elementary particle physics over at least the next 15 years as it explores the new phenomena expected to exist at the Terascale. More and more U.S. scientists and students, as well as many others from around the world, are focusing their efforts at this facility, and the United States already has made substantial contributions of resources, people, and equipment to the LHC. U.S. research groups that will carry out experiments at the LHC need to be adequately supported, and the United States should participate in upgrades of experimental facilities as those upgrades are motivated and defined through scientific results obtained from operating the facility.

Action Item 2. The United States should launch a major program of R&D, design, industrialization, and management and financing studies of the ILC accelerator and detectors.

Strong theoretical arguments and accumulating experimental results provide convincing evidence that the Terascale will provide a rich array of physics that will demand exploration by both hadron colliders (such as the LHC) and electron colliders. The consensus of the elementary particle physics community worldwide is that the ILC should be the next major experimental facility to be built. No matter what the LHC finds, an ILC will enable an even greater exploration of the mysteries of the Terascale.

The Global Design Effort (GDE) for the linear collider, which is currently under way, expects to produce an initial cost estimate based on the reference design by the end of 2006, with a full technical design proposal in 2009. An informed decision on the construction of an ILC could be made as soon as a technically credible cost estimate exists; ideally, this decision should be made no later than 2010, by which time the LHC should have revealed the nature of some of the new physics that lies at the Terascale. (The committee provides additional analysis of the path forward in Appendix A.)

Significant R&D is necessary to resolve the remaining technological challenges and to minimize the cost of this multi-billion-dollar facility. Based on evidence presented to the committee and subsequent analysis, U.S. expenditures on R&D for the ILC should be greatly increased. For the accelerator, this commitment should be as high as $100 million in the peak year, with a cumulative investment of $300 million to $500 million over the next 5 years. For the detectors, the appropriate level of resources for R&D would be perhaps $80 million over this period.

Action Item 3. The United States should announce its strong intent to become the host country for the ILC and should undertake the necessary work to provide a viable site and mount a compelling bid.

The United States should move forward in preparing a bid to host the ILC project. Such an aspiration is worthy of a great nation wishing to occupy a leadership position on the scientific and technological frontiers. Building the ILC in the United States will inspire future generations, amply repay the required investments, and lead to a much greater understanding of the universe in which we live. In addition, building and operating the ILC in the United States will provide a focal point to attract talented students and scientists from around the world to U.S. academic research institutions.

One issue that the committee did not address in its analysis was the detailed cost estimate for constructing an ILC. The committee was aware of several preliminary estimates that were developed previously in the United States and other countries, but it concluded that these estimates were based on different design concepts and did not necessarily represent the current plan for the project. The committee also has monitored closely the ongoing GDE, which is currently scheduled to produce by the end of 2006 a Reference Design Report (RDR) that will include a preliminary cost estimate based on the reference design. The committee recognizes the prudence of this approach: A credible estimate of project cost must await a specific set of design parameters and, later, international selection of a viable site. In general, the committee notes that the scale, complexity, and engineering challenges of the ILC are expected to be very roughly comparable to those associated with the LHC.

If the United States is successful in its bid to host the ILC, an increase in resources devoted to particle physics in the United States will be required. A constant-effort budget will not be sufficient to fund the U.S. share of site and mitigation costs, of housing the assembled scientific and engineering staff during construction, and of the construction and operation of the ILC accelerator and detectors.

Although site selection for the ILC will be determined through an international process, the existing physical infrastructure and human capital at Fermilab make it an advantageous site within the United States. As the only national laboratory devoted primarily to particle physics, Fermilab has an opportunity and a responsibility to the national particle physics program to secure the ILC as its top priority.

Action Item 4. Scientific priorities at the interface of particle physics, astrophysics, and cosmology should be determined through a mechanism jointly involving NSF, DOE, and NASA, with emphasis on DOE and NSF participa-

tion in projects where the intellectual and technological capabilities of particle physicists can make unique contributions. **The committee recommends that an increased share of the current U.S. elementary particle physics research budget should be allocated to the three research challenges articulated below.**

Three major research challenges in astrophysics and cosmology research could lead to discoveries with potentially momentous implications for particle physics:

- The direct detection of dark matter in terrestrial laboratories, the results of which could then be combined with measurements of candidate dark matter particles produced in accelerators.
- The precision measurement of the cosmic microwave background (CMB) polarization, which would probe the physics during the inflation that appears to have occurred within a tiny fraction of a second following the big bang.
- The measurement of key properties of dark energy.

The United States has already established itself as a leader at the interface of particle physics, astrophysics, and cosmology. Since current commitments to this area from the particle physics budgets are relatively modest compared to the full program, it is the sense of the committee that they should be built up to approximately two to three times the current level.

Action Item 5. The committee recommends that the properties of neutrinos be determined through a well-coordinated, staged program of experiments developed with international planning and cooperation.

- **A phased program of searches for the nature of neutrino mass (using neutrinoless double-beta decay) should be pursued with high priority.**
- **DOE and NSF should invite international partners in order to initiate a multiparty study to explore the feasibility of joint rather than parallel efforts in accelerator-based neutrino experiments. Major investments in this area should be evaluated in light of the outcome of this study.**
- **Longer-term goals should include experiments to unravel possible charge-parity (CP) violation in the physics of neutrinos and renewed searches for proton decay. There may be a valuable synergy between these important objectives, as the neutrino CP violation measurements might**

require a very large detector that, if placed deep underground, would also be the right instrument for detecting proton decay.

The demonstration that neutrinos have nonzero masses may be one of the first signals of the new physics expected in the years ahead, since the observed masses are in the range predicted by theoretical ideas that unify the forces of nature. In the future, neutrinoless double-beta decay experiments could demonstrate that the neutrino is its own antiparticle, which would greatly strengthen the case for interpreting neutrino masses in terms of unified theories of the fundamental forces. Furthermore, proton decay experiments might show that the proton is unstable, which would confirm one of the most basic predictions of unified theories.

Full exploitation of large, accelerator-based opportunities in neutrino physics will require planning in an international framework.

Action Item 6. U.S. participation in large-scale, high-precision experiments that probe particle physics beyond the Standard Model should continue, but the level of support that can be sustained will have to be very sensitive to the overall budget picture. Only very limited participation will be feasible in budget scenarios of little or no real growth. Participation in inexpensive, small-scale, high-precision measurements should be encouraged in any budget scenario.

The information from such studies is complementary to that obtainable via direct searches for new particles at the LHC and ILC and has historically played an important role in constraining models of new physics. Types of investigation include a future B factory, lepton-flavor violation and rare-decay studies, precision measurements of the muon g-2 parameter, and searches for electric dipole moments. Some of the latter can be relatively small-scale efforts and should be supported as part of the overall program when they offer significant reach into unexplored physics.

LOOKING TO THE FUTURE

With experimental access to the Terascale at the LHC and the proposed ILC, the particle physics community is poised for discoveries that could revolutionize how we view our world and the universe. Without question, the United States should be a leader in this great scientific adventure.

If these recommendations are carried out in accordance with the committee's strategic principles, the United States will maintain and enhance, for decades, its position as a leader in this field. Achieving these goals will require increased investment, but this investment will be richly repaid by progress across the science and technology frontier, the invigoration of particle physics, a boost in the morale of young scientists across a variety of disciplines, and the generation of new high-technology jobs.

If the United States does not win the bid for the ILC or chooses not to pursue this option, the national program still should participate vigorously in the LHC and ILC programs and expand efforts at the interface of particle physics, astrophysics, and cosmology. Without a modest budget increase, the U.S. program would have to rely on international partners to play a leading role in exploring much of the physics of the neutrino sector.

If the United States does not actively participate in exploration of the Terascale and if support for the field continues to decline, it will be clear that the United States has decided to abandon leadership in particle physics. U.S. researchers would then only be able to participate modestly in the LHC and ILC programs, and a U.S. leadership position more than half a century old would be sacrificed.

If a decision is made to host the ILC project in this country, the United States would be expected to shoulder a significant fraction of its costs. Such a course would require growth in the particle physics budget to purchase the right-of-way and to design, build, staff, and operate this forefront scientific facility.

The proposed American Competitiveness Initiative offers one way to realize many of the opportunities described in this report. By committing to a strategic vision in particle physics, the United States can remain a leader in this vital area of science and technology.

1

The Scientific Excitement
and Challenges

In 2005 the world celebrated the International Year of Physics.[1] In part, this celebration commemorated the centenary of what has become known as Albert Einstein's "miraculous year" of 1905, when he published four groundbreaking papers that laid a key part of the foundation of modern physics. It also honored other momentous discoveries in physics of the past century, including the development of quantum mechanics and the successful testing of what is known as the Standard Model of elementary particle physics—advances that have led to a new understanding of nature and to technologies that have profoundly influenced our lives.

In the sciences in general, the hundred years between 1905 and 2005 eventually could become known as the "miraculous century." Greater understanding of the constituents and properties of materials resulted in an unprecedented array of new products and industrial processes. The discovery of the structure and function of DNA deepened our understanding of genetic inheritance and human development and gave researchers the ability to alter the genetic material of living organisms. The discovery of plate tectonics contributed to a new view of Earth as an integrated biological and physical system in which humans are playing an increasing role. In short, advances throughout the sciences during the 20th century revealed many of nature's secrets and radically changed our view of the world.

In physics in particular, the advances of the 20th century were unprecedented.

---

[1] For additional information, see <http://www.physics2005.org/>.

One of Einstein's 1905 papers described the special theory of relativity, which explained that moving objects become more massive as they approach the speed of light, clocks slow down, and objects flatten into pancakes. In 1916, Einstein published his general theory of relativity, showing that mass warps the structure of space and time, accelerating objects emit gravitational waves, and clocks slow down in a gravitational field. In the 1920s and 1930s, physicists developed the set of ideas known as quantum mechanics to explain the puzzling behavior of the subatomic world; these fundamental insights contributed to some of the most important technologies of the 20th century, including the semiconductors that have made possible the proliferation of modern electronic devices. Also in the 1920s and 1930s, astronomers produced evidence indicating that the universe is expanding, which suggests that all matter was created in an event known as the big bang, which took place more than 13 billion years ago. Studies of materials revealed new phenomena such as superconductivity, nuclear fission, and the coherent emission of light (leading to the development of the laser). These astonishing insights into the nature of the physical world produced new fields of physics (such as nuclear physics, condensed matter physics, and particle physics), generated knowledge that found applications throughout the sciences and in technology, and created a base of understanding that has helped remake our world.

The field of elementary particle physics (or, simply, "particle physics," which is the term used most often in this report) took shape in the first half of the 20th century as physicists began to study the fundamental constituents of matter and their interactions (Box 1-1). Both experimentation and theory have been critical to the advance of particle physics. For example, early in the 20th century, certain puzzling experimental results caused physicists to seek new and more fundamental explanations of the laws of nature. This search led to Einstein's startling new theories of space and time and of gravity, as well as to the equally revolutionary development of quantum mechanics by physicists such as Max Planck, Niels Bohr, Werner Heisenberg, Max Born, and Erwin Schrödinger. The second half of the century witnessed a blossoming of particle physics as experiments tested existing hypotheses and inspired new ones. Many of those experiments involved particle accelerators, which convert matter to energy and back to matter again, as described by Einstein's equation, $E = mc^2$. In recent decades, accelerator experiments have become enormous undertakings involving thousands of scientists and engineers and intellectual and financial contributions from countries around the world. In addition, a spectrum of much smaller, less expensive, but also highly valuable experiments has measured the special properties of particles and particular interactions among particles. Most recently, astronomical data from satellites and ground-based facilities have produced extremely useful information for particle physics. The nascent field of particle astrophysics has brought a deeper apprecia-

---

BOX 1-1
What Is Elementary Particle Physics?

Physics has demonstrated that the everyday phenomena we experience are governed by universal principles applying at time and distance scales far beyond normal human experience. Elementary particle physics is one avenue of scientific inquiry into these principles. What rules govern energy, matter, space, and time at the most elementary levels? How are phenomena at the smallest and largest scales of time and distance connected?

To address these questions, particle physicists seek to isolate, create, and identify elementary interactions of the most basic constituents of the universe. One approach is to create a beam of elementary particles in an accelerator and to study the behavior of those particles—for instance, when they impinge upon a piece of material or when they collide with another beam of particles. Other experiments exploit naturally occurring particles, including those created in the sun or resulting from cosmic rays striking Earth's atmosphere. Some experiments involve studying ordinary materials in large quantities to discern rare phenomena or search for as-yet-unseen phenomena. All of these experiments rely on sophisticated detectors that employ a range of advanced technologies to measure and record particle properties.

Particle physicists also use results from ground- and space-based telescopes to study the elementary particles and the forces that govern their interactions. This category of experiments highlights the increasing importance of the intersection of particle physics, astronomy, astrophysics, and cosmology. In general, large, centralized infrastructure, such as large accelerators, telescopes, and detectors, plays a crucial role in enabling particle physics. Working together in large teams, particle physicists construct and operate these complex facilities and then share the results. Not all experiments are so large, however, and progress in particle physics depends on the combined efforts of large and small projects.

---

tion of the fundamental connection between the study of elementary particles and such astronomical phenomena as active galactic nuclei, black holes, pulsars, and the overall evolution of the universe.

Over the entire suite of experiments and observations spreads the umbrella of theory. Theoretical physicists seek to construct a coherent intellectual edifice that can encompass and explain what has been seen, using the power of mathematics to make their ideas precise and logically consistent. From these theoretical models emerge predictions that help define the critical experiments needed to test the current framework and extend today's understanding to new phenomena.

This sustained real-time interplay of experiment and theory has produced astonishing progress. In the first part of the 20th century, physicists learned that all matter here on Earth is built out of subatomic particles known as electrons, protons, and neutrons. In the second half of the century, they discovered that protons and neutrons are composed of more fundamental particles known as quarks, and that the quarks and electrons that constitute everyday matter belong to families that include heavier and much rarer particles. They learned that particles interact through just four forces: gravity, electromagnetism, and two less familiar forces known as the strong force and the weak force. They developed a theoretical framework known as the Standard Model, which describes and predicts the behavior of elementary particles with extremely high levels of precision. The development and extraordinarily precise testing of the Standard Model have been among the crowning achievements of 20th century science.

Yet considerable evidence suggests that the advances of the 20th century rather than ending the story have set the stage for a new era of equally exciting progress. Results from both experiment and theory suggest that the next few decades will produce information that could help answer some of the most basic questions scientists can ask: Why do particles have mass? What are the relationships between the forces observed in nature? What accounts for the structure and evolution of the universe, and what is its future?

These questions are ripe for a new phase of investigation for a range of reasons. For decades, physicists have had strong reasons to think that great discoveries await experiments that can be conducted at what is known as the Terascale. "Tera" refers to the million million electron volts of energy that can be imparted to particles in the most powerful accelerators available. It has taken more than 75 years to develop the technologies needed to construct accelerators that can open this new frontier. At last, experimental facilities are being constructed that bring the Terascale within reach. Other experiments examining high-energy cosmic rays generated in the distant universe or neutrinos generated by solar fusion also promise to complement in extremely valuable ways the information generated by accelerators.

Promising experiments currently under way at Fermi National Accelerator Laboratory (Fermilab) have begun to explore the lower reaches of the Terascale. In 2007 the Large Hadron Collider (LHC) at the European Center for Nuclear Research (CERN) is scheduled to begin colliding protons. This facility will for the first time provide physicists with the ability to carry out controlled laboratory studies at a broad range of energy levels within the Terascale range. Moreover, the prospect of further exploiting the Terascale with a new accelerator known as the International Linear Collider (ILC) has galvanized particle physicists from around the world to consider in detail how currently available technologies could be used to address compelling scientific questions beyond the reach of the LHC alone.

CHALLENGES TO THE STANDARD MODEL

Why is the Terascale so important?

At the Terascale, two of the main forces in nature, the weak and electromagnetic forces, appear to join together to become a single entity. Exactly how this happens is a mystery. There is a proposal within the framework of the Standard

---

BOX 1-2
Einstein's Dream

After Albert Einstein published his general theory of relativity in 1916, he devoted much of his scientific work to a problem that consumed him until the end of his life in 1955: the unification of the fundamental forces of nature, including electromagnetism, gravity, and the forces active within the atomic nucleus. Einstein's dream was to develop a unified field theory that would describe in a single set of equations all the seemingly distinct forces that act on particles. Though he worked on the problem until the day he died, he never solved it.

Today physicists still have not achieved a unified theory of the fundamental forces. But new theoretical ideas and experimental results have resulted in extremely promising hypotheses. The discovery of phenomena unknown to Einstein, such as quarks, dark matter, and dark energy, means that physicists may be on the verge of realizing Einstein's goal. The next generation of experimental facilities may bring Einstein's dream within reach.

Model, but it has never been tested and it raises baffling theoretical questions. Understanding how the weak and electromagnetic forces are unified is believed to be an important part of understanding the broader unification of particle forces, perhaps including gravity, in keeping with Einstein's aesthetic dream of unifying all the laws of nature (see Box 1-2).

How the weak and electromagnetic forces are unified is a question that can only be answered using accelerators. For example, it is not possible to make these measurements using cosmic rays, because the highest energy cosmic rays are too few and it is not possible to study them with enough precision.

Scientists everywhere seek the simplest possible explanation of the phenomena they study that will survive scientific scrutiny. In physics, the development of a single coherent scientific framework that would explain the nature of matter, its mass, its evolution, and the forces associated with it has inspired the work and dreams of generations of physicists. Moreover, the scientific unification of seemingly diverse phenomena often generates great intellectual dividends, as occurred with the unification of electricity and magnetism in the 19th century. The next important step in this program of unification requires the direct investigation of the Terascale.

Both theory and past experiments strongly indicate that new phenomena await discovery in this energy range. A world of new particles predicted by a hypothesis known as supersymmetry may be seen, and these new particles could provide essential information about already known particles. The particles that constitute the dark matter responsible for the formation of galaxies may appear at these energies. The Terascale may be the gateway to new dimensions of space beyond those we experience directly but that nevertheless can have an important impact on our world. New phenomena appearing at the Terascale could include a particle called the Higgs boson, which is responsible for the mass of the known particles. Or, these new phenomena could take an entirely different form, including phenomena that are completely unexpected and not yet imagined. All of these possibilities can best be explored at accelerators.

Exploring Terascale physics is the essential next step in addressing the most exciting scientific challenges in particle physics. Particle physics appears to be on the verge of one of the most exciting periods in its history.

The Standard Model provides an excellent and carefully tested description of the subatomic world at the energy levels that currently can be studied in laboratories. However, at energy levels that physicists are only now beginning to access experimentally, the Standard Model is incomplete. This strongly suggests that exciting new discoveries loom in the years immediately ahead, especially as the LHC begins to probe this energy region. It also suggests that these impending discoveries may transform our understanding of the origin of matter and energy and the ongoing evolution of the universe.

The limitations of the Standard Model are evident, for example, when trying to account for the force of gravity. The Standard Model incorporates the forces of electromagnetism and the strong and weak forces. But when physicists attempt to include gravity as a fourth force in the Standard Model, they run into severe mathematical inconsistencies. Thus, two pillars of 20th century physics—gravity (as described by Einstein's general theory of relativity) and quantum mechanics—require some new theoretical framework that can include them both.

Astronomical discoveries pose another severe challenge to the Standard Model. Astronomical observations have shown that protons, neutrons, electrons, and photons—which account for everything with which we are familiar—make up less than 4 percent of the total mass and energy in the universe. About 20 percent consists of some form of dark matter: massive particles or conglomerations of particles that do not shine and do not scatter or absorb light. Astronomers can detect dark matter by observing how it distorts the images of distant galaxies, an effect known as gravitational lensing, and they can map the distribution of dark matter throughout space. The composition of dark matter is not yet known; it may consist of a cloud of elementary particles of some unknown sort, though there are other possibilities. Yet we owe our existence to dark matter. Without the added gravitational attraction of dark matter, the stars and galaxies, including our own Milky Way, would likely never have formed, because the expansion of the universe would have dispersed the ordinary matter too quickly.

More surprising still is the fact that most of the energy of the universe today consists of something else entirely—an ephemeral dark energy that gravitationally repels itself. A clump of ordinary matter or dark matter has an attractive gravitational force that draws matter together and slows down the expansion of the universe, but dark energy pushes itself apart and acts to speed up the expansion of the cosmos. Because most of the energy in the universe is dark energy, the expansion of the universe is accelerating. Thus, dark matter played a crucial role in the past by causing galaxies to form, and dark energy will play a crucial role in the continuing evolution of the universe. What dark matter and dark energy are and how they fit into the overall understanding of matter, energy, space, and time are among the most compelling scientific questions of our time.

The predominance of matter over antimatter in the universe also poses problems for the Standard Model. In 1928, Dirac's incorporation of Einstein's special theory of relativity into quantum mechanics suggested that, for each kind of elementary particle, there is an antiparticle with the same mass and opposite charge. When a particle and its antiparticle come together, they are both annihilated and their mass is converted into radiant energy. Experiments using antimatter in high-energy physics laboratories show that the fundamental forces act nearly the same on particles and antiparticles except for small differences that can be explained using the Standard Model. However, the Standard Model cannot explain why the universe consists almost entirely of matter and almost no antimatter. This asymmetry is a good thing, since otherwise so much matter and antimatter would have been annihilated in the early universe that there would not have been enough to make stars and planets. Yet the cause of the large imbalance is a mystery. Many physicists believe that the imbalance was created by physical processes that occurred as the universe was cooling after the big bang. It may be possible to study some of the same physical processes by colliding elementary particles at high energies in accelerators.

Another outstanding question involves the early evolution of the universe. Most cosmologists believe that the large-scale structure of the universe was created by a burst of "inflation," a brief period of hyperaccelerated expansion during the first 10^{-30} second after the big bang, perhaps associated with interactions involving dark energy. This inflation could have rapidly smoothed out the distribution of matter and energy except for tiny lumps here and there that later became the seeds for galaxy formation. Recent observations of the cosmic background radiation have provided exquisitely precise corroborating evidence for this picture of inflation, but there remains a key missing component—the explanation for what drove the hyper-expansion. The Standard Model does not provide an answer, but new physical laws discovered using the next generation of high-energy accelerators may provide essential clues.

New evidence about the properties of the elusive particles known as neutrinos also raises exciting new questions. Neutrinos are extremely numerous in the universe but interact very rarely with the basic constituents of ordinary matter—literally billions and billions of neutrinos pass unaltered through each of us every second. A beautiful series of experiments has demonstrated that neutrinos, long thought to be without mass, instead have very small masses—approximately 1/200,000th the mass of the electron, which already has an extremely small mass by subatomic standards. Moreover, the neutrinos produced in nature are apparently not in states of definite mass. This phenomenon, which would baffle a classical physicist, is a typical effect of quantum mechanics. It has a peculiar consequence: Neutrinos can spontaneously change from one type to another, an effect known as "neutrino oscillations." Neutrino masses do not fit into the Standard Model, so these new observations have necessitated the first major extension to the model in three decades. Exactly what further extensions are required will not be known until the completion of currently operating neutrino experiments as well as the next generation of experiments that are now being planned and initiated.

Thus, at the start of the 21st century, particle physics experiments, astronomical observations, and theoretical developments in both particle physics and cosmology point to exciting new phenomena that are just on the verge of being observed. Combining quantum theory and general relativity, and understanding dark matter and dark energy, will require new ideas and new experiments. The technologies needed to conduct these experiments are now available. As a result, particle physics is poised on the brink of a scientific revolution as profound as the one Einstein and others ushered in early in the 20th century. There is every possibility that these Tersacale discoveries will have an equally important impact across the fields of science.

RESPONDING TO THE CHALLENGES

Physicists use a variety of natural phenomena to study elementary particles and their interactions. Extremely energetic particles are created in the distant cosmos and stream to Earth as cosmic rays, where they can be observed in special detectors. Studies of neutrinos generated within the sun were critical in establishing that neutrinos have mass. Nuclear reactors are sources of intense flows of neutrinos. Physicists will continue to observe and study these particles in a variety of laboratories, including laboratories embedded in ice or deep underground.

However, most of the particles that physicists study are created in particle accelerators and observed in specialized detectors located at domestic laboratories and at laboratories in other countries. Such accelerators convert energy into particles that were abundant shortly after the big bang but are extremely rare today; accelerators also provide a window onto interactions among particles that are apparent only at high energies. Studying these particles under controlled laboratory conditions has been, and will continue to be, essential to understanding topics ranging from the origins of matter to the nature of the universe. In particular, comprehensive exploration of the Terascale will require the use of accelerators to elucidate nature's underlying physical principles.

The most powerful accelerator in existence today is the Tevatron at Fermi National Accelerator Laboratory outside Chicago. Before the end of the decade, when it is scheduled to be shut down, the Tevatron will explore the lower reaches of the Terascale and may make important new discoveries about the Higgs boson and the possible existence of new particles predicted in some extensions of the Standard Model.

However, the next major set of discoveries is likely to come from a very exciting set of experiments at a new accelerator, the LHC in Geneva, which is scheduled to begin operating in 2007. This machine will enable physicists to explore energy regions inaccessible to Fermilab's Tevatron. The LHC is a project of CERN, the international laboratory established in 1954 as a joint venture of 12 European countries; CERN currently has 20 member states, all in Europe. The LHC will make CERN the most important center in the world for particle physics over the next decade. The United States has participated both in building the accelerator and in the large collaborations that are building the detectors. U.S. participation has been an important contributor to this tremendous scientific opportunity.

The experimental facilities required to reach the Terascale and record the necessary data are exceedingly complex and costly. As the activities at CERN have demonstrated, some of the most advanced experimental facilities, especially those exploring the energy frontier under controlled conditions, are beyond the resources that any single country, or even a single region of the world, can be expected to commit to particle physics. Moreover, these technologically complex facilities require the contributions of many scientists and engineers from throughout the world with different mixes of skills. These factors have caused experimental particle physics to become a truly international activity. No matter what future program of particle physics the United States supports, international collaborations of various kinds will become more essential than ever to the advance of particle physics and to the vitality of the U.S. program in particle physics.

In one sense, all of science is becoming increasingly internationalized. New information flows easily and quickly around the world and is shared, almost in real time, with interested scientists wherever they are located. Such information flows also characterize the world of particle physics. However, particle physicists also need to assemble geographically, often in international teams, at national or regional laboratories to jointly plan and carry out particular experiments. Moreover, such experiments typically take 5 to 10 years or more from the initial set of ideas to the full analysis of the results. As a result, the field of particle physics has developed its own distinctive sociology, which is characterized by a great deal of movement of scientists, engineers, and students across international borders and a full acceptance of the interdependence of the scientific world. The capacity to welcome scientists from abroad as full partners, wherever the key experimental facilities are located, is an essential requirement for the field of particle physics. No nation or region can provide all the experimental facilities to meet the full needs and interests of its community of particle physicists; as a result, international partnerships of various kinds have been developed to solve this problem.[2]

THE ROLE OF THE UNITED STATES IN PARTICLE PHYSICS

For the last 50 years the United States has been at the forefront of particle physics. That leadership position has had an immense impact on this country. It has inspired generations of young people to become members of the strongest scientific workforce in the world. It has attracted outstanding scientists from abroad to come to the United States and contribute to the nation's intellectual and economic vitality. The novel technologies developed to carry out particle physics experiments have had widespread applications in other areas of science and industry (see Box 1-3).

Despite its historic accomplishments, the U.S. program in particle physics is at a crossroads. Great scientific opportunities lie immediately ahead, but the challenge of mobilizing the U.S. program to exploit this special moment is significant. In fact, it is not at all clear that the United States can continue to occupy a leadership position in the worldwide particle physics community. There are several reasons for this situation. First, despite the growing sense of scientific excitement and opportunity within particle physics, and despite a decade of strong national economic growth, no additional resources have been devoted in recent years to the U.S. program in particle physics (see Box 1-4). This stand-still budget contrasts strongly with the situation abroad, where Europe and Japan are both making new commitments to take advantage of exciting scientific opportunities.[3] In addition,

---

[2]The emerging dominance of large-scale facilities for science has attracted considerable attention. In a 2003 report, the German Science Council made the following observations: (1) the success of basic research in natural sciences often is based on the use of complex and costly large facilities, (2) large facilities should be initiated by a broad scientific user community, and (3) a central role of the operation of a large facility lies in the close connection of top-level scientific research and the education of young scientists and their integration into international research collaborations. See German Science Council, *Theses on the Significance of Large-Scale Facilities for Basic Scientific Research*, 2003.

[3]It is the committee's rough estimate that Europe, through local spending by member states and through the collective national funding of CERN, invests about twice as much each year in particle physics as does the United States, and that Japan alone invests about half as much as does the United States.

BOX 1-3
Particle Physics in Science and Society

The world's most powerful accelerators, which are among the largest and most technologically sophisticated experimental devices ever built, are tremendously impressive machines that involve remarkable feats of engineering. They have also generated waves of technological innovations and applications throughout the sciences and society.

One notable example in recent years was the development of the key protocols that underpin the World Wide Web. Building on the backbone of the already existing Internet, this new way of sharing information has revolutionized the way the world communicates and does business. These protocols were initially developed by a researcher at CERN seeking better ways for large groups of particle physicists to share information and collaborate on experiments.

The small accelerators used in hospitals to generate x rays for radiation treatment come from designs developed for particle physics. These designs have been improved and refined as research on accelerator technologies for forefront science continues to be applied to medical accelerators. Roughly 100,000 patients are treated every day in the United States with radiation from electron beam accelerators. Accelerators also are used to produce radioisotopes for treatment, diagnostic tools, and research, and technologies developed for detecting particles in high-energy physics experiments have had important applications in medical imaging.

When energetic charged particles pass through curved paths in a magnetic field, they generate radiation. The ability of accelerators to produce powerful beams of x rays or photons of differing energies has generated applications across a broad range of science. Each year as many as 40,000 U.S. researchers from many different scientific disciplines use these powerful light beams to conduct experiments. Accelerator x-ray sources provide, for example, the ability to decipher the structure of proteins and other biological macromolecules and to find trace impurities in the environment or on the surface of a silicon chip. The science produced by these experiments has found applications throughout industry and medicine.

In general, particle physics contributes to—and depends on—advances in other areas of physics (such as nuclear physics and condensed matter physics) and in many other scientific fields, including materials science, computing, biology, chemistry, and nanoscience. The health of science requires support of all parts of this interlocking web.

Technical challenges faced by particle physicists—such as processing millions of signals quickly, using distributed computers to solve complex problems, and generating electromagnetic fields to accelerate and confine charged particles—have led to many spinoff technologies. Particle physics also has contributed in important ways to mathematics, even as mathematics has been used to understand the theoretical structures describing particles.

In industry, accelerators are used for R&D, manufacturing, testing, and process control. For example, beams from accelerators are used to alter the composition of materials and to improve the characteristics of products. Uses of accelerators range from the dating of archaeological samples to the simulation of cosmic rays to determine the impact of radiation on space-based electronics.

Finally, because particle physics addresses some of the deepest questions that humans can ask, it resonates strongly with the public at large. The science shelves of bookstores teem with popular expositions of the current understanding of these issues, and many students are attracted to science because they are interested in issues addressed by particle physics.

BOX 1-4
Federal Investments in Particle Physics over the Past Decade

Two federal agencies have been the main source of funding for elementary particle physics: the Office of High Energy Physics in the Department of Energy's (DOE's) Office of Science and the Physics Division in the National Science Foundation's (NSF's) Directorate for Mathematical and Physical Sciences. Although support from the NSF plays a crucial role, it is DOE that provides by far the largest share of resources for particle physics and maintains the key national laboratories (see Figure 1-4-1). In addition, the National Aeronautic and Space Administration (NASA) and NSF provide important support for a range of astronomical experiments that are related to particle physics.

Support from DOE for particle physics has averaged about $720 million per year over the last 5 years. This funding supported the nation's flagship accelerator facilities at the Stanford

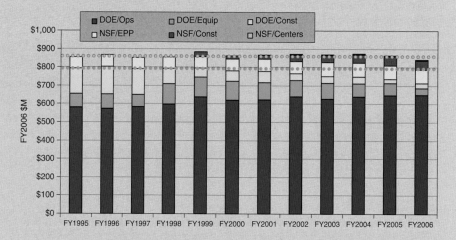

FIGURE 1-4-1 Federal investments in elementary particle physics research at DOE and NSF have remained relatively constant in inflation-adjusted FY2006 dollars. The lower dashed line extends from the FY1995 level of DOE investment to the current levels (projected) for FY2006. The upper dashed line extends from the FY1995 total federal investment to FY2006 and indicates that, as a whole, the federal portfolio has maintained a nearly constant effort over the past decade. The NSF construction investment is dominated by the building of the IceCube neutrino observatory. Note that new construction initiatives have decreased significantly and that DOE operations costs have become a larger percentage of the total.

Linear Accelerator Center (SLAC) and the Fermi National Accelerator Laboratory (Fermilab) (see Figure 1-4-2). Smaller accelerators have operated at other national laboratory and university sites, and many other universities and research organizations have programs in particle physics supported by the DOE and NSF. The direct contribution from NSF, including construction funding, has ranged from $60 million to $125 million per year over the last decade.

For the past 10 years, the nation's investments in elementary particle physics have remained relatively constant in inflation-adjusted dollars. However, this flat budget has been achieved only because declining support from DOE has been counterbalanced by what may be a short-term increase in NSF support, particularly for the construction of the large antarctic neutrino observatory IceCube. The President's proposed FY2007 American Competitiveness Initiative, with its increase in real funding for particle physics (about 8 percent), could enable many of the exciting opportunities described in this report.

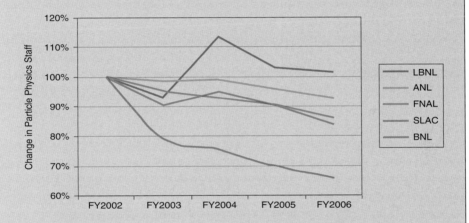

FIGURE 1-4-2 Laboratory scientific staff working in particle physics at Argonne National Laboratory (ANL), Fermi National Accelerator Laboratory (Fermilab or FNAL), Stanford Linear Accelerator Center (SLAC), and Brookhaven National Laboratory (BNL) have all declined in recent years (measured in full-time-equivalent employees), with the decrease at BNL from FY2002 to FY2003 being driven partly by the termination of the high-energy physics program of the Alternating Gradient Synchrotron (AGS). Only at Lawrence Berkeley National Laboratory (LBNL) did the staff grow from FY2003 to FY2004, owing to work on the proposed Supernova/Acceleration Probe (SNAP), a satellite designed to study dark energy through the discovery and measurement of thousands of distant supernovas.

the investments required to construct new facilities or to operate existing ones effectively have grown because of the increasing sophistication of accelerator and detector technologies.

Second, the future of experimental facilities for particle physics in the United States is uncertain. Several of the country's flagship particle physics experiments are scheduled to be shut down within the next few years, and at least one major facility (the accelerator at the Stanford Linear Accelerator Center in California) is being redirected toward other scientific areas. Some new small-scale projects related to particle physics are under construction, but no approved plans are in place for new initiatives in the United States that could capitalize on the exciting scientific challenges that have recently crystallized. Again, in other regions of the world—especially in Europe and Asia—programs in particle physics are being expanded and new experimental facilities are under construction.

Third, there has been insufficient investment in R&D for the tools and facilities of particle physics—namely, accelerators and detectors. More generally, not enough resources have been deployed in the United States to carry out smaller scale but critical experiments or to undertake critical initial explorations of the new technological frontiers for accelerators and detectors that would help enable future experiments. To continue to explore the energy frontier, more efficient and less expensive technologies are essential, and smaller scale science will remain a critical component of the national program.

Finally—and most important—since the cancellation of the Superconducting Super Collider (SSC) in 1993, the U.S. program in particle physics has not had an overall strategic plan informed by a long-term vision that would give it a unique shape, direction, and excitement. In retrospect, many feel that the failure to proceed with the SSC was a lost opportunity both for the U.S. particle physics program and for the entire U.S. scientific enterprise. Moreover, it delayed scientific progress in particle physics by at least a decade.

To grasp the opportunities now available in particle physics, a new vision is needed—a vision that can mobilize the creativity, excitement, and leadership of students and scientists and generate the public commitment needed to maintain U.S. scientific and technological leadership in particle physics.

The United States, despite not being a member state of CERN, has made and continues to make substantial commitments of both intellectual and financial resources to the LHC. Many U.S. students and scientists are participating in the upcoming experiments at the LHC. Within the next few years, more than half of U.S. experimental particle physicists will be focused on experiments occurring at CERN rather than in the United States.

If the United States maintains its present course and chooses not to seize the opportunity to explore the Terascale with distinctive U.S.-based, next-generation experimental facilities, leadership in particle physics will move to Europe, Japan, and elsewhere. Indeed, such a migration of leadership has already begun. The committee believes that if an exciting forefront particle accelerator is not available in the United States in coming decades, fewer of our young people will be attracted to particle physics and to science in general. Without such a facility, U.S. scientists and engineers will have to travel abroad to work on forefront accelerator experiments, as has already begun happening with the LHC. Universities will be increasingly reluctant to appoint particle physicists to their faculties, knowing that they will have to spend large portions of their careers working abroad. Leading scientists and engineers from other countries will no longer travel to the United States in large numbers to participate in high-energy accelerator experiments, where in the past they contributed in so many ways to the nation's scientific, cultural, and economic vitality. Eventually, particle physics in the United States will lose its vitality, with most of the important advances occurring in other parts of the world.

The committee has concluded that the price the United States would pay by forfeiting a leadership position in particle physics is too high.[4] Leadership in science remains central to the economic and cultural vitality of the United States.[5] To fuel the innovation economy of the 21st century, to maintain national security, and to produce the knowledge needed to ensure our well-being in the face of an uncertain and challenging world, the United States needs more than ever to have a strong base of science and technology. A strong scientific enterprise attracts ambitious and talented students to science. It also makes the United States a desirable place for excellent scientists from abroad to pursue some of the most important challenges on the scientific frontier. Particle physics contributes greatly to the strength of U.S. science and technology while allowing U.S. students and scientists and engineers to participate in and benefit from a worldwide scientific activity. More generally, leadership in particle physics can serve, as it has in the past, as an important symbol of leadership in science and technology.

In the flat world that is taking shape, leadership in particle physics no longer consists of single-handed efforts to maintain dominance in a particular subfield. Rather, leadership emerges from the creativity and initiative needed to organize international teams of collaborators to pursue projects that are beyond the capability of any one country. Such leadership requires making investments both at home and abroad in order to participate in and benefit from developments across

---

[4]House Committee on Science, *Unlocking Our Future: Toward a New National Science Policy*, September 1998. Available online at <http://www.house.gov/science/science_policy_report.htm>.

[5]NAS, NAE, and IOM, *Rising Above the Gathering Storm: Energizing and Employing America for a Brighter Economic Future*, Washington, D.C.: The National Academies Press, 2005 (Prepublication).

the broad scientific and technology frontier. When such investments are joint ventures with colleagues from abroad, all partners participate directly in both the costs and benefits of the enterprise. However, the capacity to deploy new discoveries across a broad spectrum of economic activities depends on the structure, incentives, and capacity of individual economies to adapt and encourage change. Careful studies in the United States indicate that investments in high-quality science and scientific leadership repay those investments many times over.[6]

Chapter 2 reviews the most important and exciting scientific questions in particle physics within the context of some of the significant milestones in the development of particle physics between 1950 and the present. Chapter 3 discusses the types of experimental facilities, scientific approaches, and devices that will be needed to explore the questions posed in the preceding chapter. In addition, it describes the evolving international framework within which decisions in particle physics must be made. Chapter 4 highlights the strategic framework within which the committee believes decisions on priorities for the future of particle physics should be made. Chapter 5 focuses on the findings and recommended action items of the committee.

Particle physics is a discovery-based science that probes the deep secrets of nature. What are the characteristics of space and time? How did the universe evolve, and how will it evolve into the future? Is nature understandable, or are there fundamental limits to knowledge? These are questions that capture the imagination of people everywhere and that great nations should strive to answer. Indeed, the nations that lead the way in answering these questions will occupy a special place in human history.

If the United States is to continue to be a leader in particle physics, it must provide appropriate support for scientists and students working at the scientific and technological frontiers in particle physics, leverage resources by pursuing joint efforts with international partners, and—above all—adopt a strategic framework and an associated set of priorities to maximize the impact of the resources that are available. The administration's proposed budget for FY2007 takes up this challenge and begins to provide the necessary resources for the physical sciences and mathematics to sustain their vitality and the vitality of U.S. science. This report charts a path toward the future for particle physics that will make the field's tremendous potential a reality.

---

[6]See, for example, NAE, *The Impact of Academic Research on Industrial Performance*, Washington, D.C.: The National Academies Press, 2003; Council on Competitiveness, *Innovate America: Thriving in a World of Challenge and Change*, Washington, D.C.: Council on Competitiveness, 2004.

2

Key Questions in Particle Physics

P article physics had relatively simple origins, beginning with the study of natural sources of particles, either radioactive atoms or cosmic rays from space. As one discovery led to another, surprises proliferated. New questions emerged, and newer and more powerful instruments were developed to answer them.

Now particle physics has advanced to the point that it can ask some very deep questions:

- Can all the forces between particles be understood in a unified framework?
- What do the properties of particles reveal about the nature and origin of matter and the properties of space and time?
- What are dark matter and dark energy, and how has quantum mechanics influenced the structure of the universe?

Some of these questions have a long history going back to the earliest days of particle physics and even before that. Some of them are new questions raised by contemporary discoveries. What the questions have in common is that progress in experiment and theory has revealed new clues and created fundamentally new ways of answering them. The committee deals with each of these key questions in turn.[1]

---

[1]This chapter provides a short background on the historical development (since about 1950) of some key themes in elementary particle physics.

CAN ALL THE FORCES BETWEEN PARTICLES BE UNDERSTOOD IN A UNIFIED FRAMEWORK?

Even in preindustrial times, people knew about static electricity, lodestones (or magnetized rocks), and light. From a modern point of view, this means that one of the fundamental forces of nature—electromagnetism—was observed without any modern technology. Of course, preindustrial people did not know that static electricity, magnetism, and light are different aspects of the same thing. This only became clear when James Clerk Maxwell combined electric and magnetic forces into the theory of electromagnetism in the mid-19th century. Maxwell's equations—together with the discovery of the first elementary particle, the electron, in 1897—led to the invention of radio and, ultimately, to today's electronic technologies.

One other fundamental force was known before the 20th century—gravity. Gravity is vastly weaker than the other forces—so weak that the gravitational forces between individual elementary particles are too small to observe. Yet the gravitational effects of many particles are cumulative. Thus for everyday objects gravity is clearly observable, and gravity is the dominant force for galaxies and in the universe as a whole today.

The advanced technology of the 20th century was required to discover and understand the two other forces that influence the behavior of particles. Some atoms decay radioactively by emitting electrons and neutrinos. In the 20th century these decays were shown to be the product of weak force interactions. The weak force—which is critically important in stellar processes, the formation of the elements beyond iron, and the evolution of the early universe—is just as fundamental as electromagnetism or gravity, but it is far less obvious in everyday experience.

Recognition of the strong, or nuclear, force resulted from research into the atomic nucleus. The nucleus consists of protons and neutrons that are bound together in a tiny ball. Protons have a positive electric charge, which makes them repel each other. However, something keeps the nucleus from flying apart. This something is the strong force.

Understanding the strong and weak forces depends centrally on quantum mechanics. In the 1920s, physicists began studying the properties and behaviors of particles, in part to understand the forces between them. This process culminated a half century later with the emergence of the Standard Model. The Standard Model, in a remarkably concise way, describes and explains many of the phenomena that underlie particle physics and captures with astonishing precision an incredible range of observational data.

The Standard Model has another important feature. It reveals a deep analogy between the four forces, in keeping with Einstein's goal of unifying all of the fundamental forces. All are described by similar equations. In the Standard Model, the electromagnetic force, weak interactions, and the strong force are described by equations called the Yang-Mills equations, which generalize Maxwell's equations of electromagnetism. These Yang-Mills equations have a close analogy with Einstein's equations of gravity in his general theory of relativity. Understanding the similarities and differences among these forces and their mathematical representations will be a key to realizing Einstein's dream.

In the Standard Model, each force is carried by a different kind of particle. That is, forces are exerted by the exchange of certain particles between two objects. The photon, which is the basic quantum unit of light, carries the electromagnetic force. The weak force is carried by particles known as W and Z bosons. The strong force, now understood as the force that binds quarks to form particles such as protons and neutrons, is carried by particles known as gluons. Like quarks, gluons are not seen in isolation because of the strength of the forces binding them together. The gluons, therefore, must be observed indirectly, by the patterns of particle production that they cause in high-energy experiments. These patterns have been studied, and the results match the theory over a wide range of energies.

According to the Standard Model, electromagnetism and the weak force have a related origin, which is why the two are sometimes described as electroweak interactions. Electromagnetism is mediated by photons that obey Maxwell's equations. Weak interactions are mediated by W and Z particles that obey the analogous Yang-Mills equations. The W and Z bosons have a very large mass—nearly one hundred times the mass of a proton. Why are the masses of the W and Z particles so large, whereas the photon has no mass? Why are the force-carrying particles so different, with the photon being detectable by our eyes while the W and Z particles can be observed only with the most sophisticated equipment? Settling this question, which would explain why the weak interactions are weak, is a major goal of particle physics for the coming decade.

To put the question differently, if the equations are so similar, why are the forces so different? According to the Standard Model, the mechanism for breaking the symmetry between the two forces is something called "spontaneous symmetry breaking" (see Box 2-1). Exactly how this symmetry breaking occurs remains unknown. This process determines which particle of the three (the photon, the W particle, the Z particle) remains massless while the others become massive. Furthermore, the theory predicts that there must be at least one more particle associated with the symmetry breaking. In the Standard Model, there is a single such particle: the Higgs boson. The field associated with this particle gives mass to matter by acting as a kind of invisible quantum liquid that fills the universe. Interactions with this quantum liquid give all particles mass. Heavier objects, such as the W and Z particles, are more strongly affected by the Higgs field, lighter ones

---

BOX 2-1
Symmetry Breaking

One of the most important concepts in physics is spontaneous symmetry breaking. The laws of nature often have a good deal more symmetry than the phenomena that we actually observe. The reason is that the lowest energy state of a system often does not have the full symmetry inherent in the laws. An example is provided by a ball placed at the top of a sombrero, as in Figure 2-1-1.

FIGURE 2-1-1 An example of spontaneous symmetry breaking.

---

interact less with it, and massless particles like the photon slip through the field without feeling it at all.

The Higgs particle, which is the particle associated with the Higgs field, has not yet been seen. One major goal of upcoming accelerator experiments is to discover whether a simple Higgs particle causes the breaking of the symmetry between the weak interactions and electromagnetism, as in the Standard Model, or whether there is some more complicated mechanism. The mass of the Higgs particle (or whatever breaks the electroweak symmetry) can be roughly estimated. The masses of the W and Z particles are 80 and 91 GeV. (GeV refers to giga-electron volts, which is a way of describing the mass of a particle in terms of its energy equivalent; 1 GeV is approximately the mass of a proton, and 1,000 GeV equals 1 TeV.) Existing accelerators would have observed the Higgs particle if its mass had been less than 115 GeV and if it decayed as predicted by the Standard Model. Since it has not been observed, it must be more massive than that. However, the Standard

When the ball sits on top of the hill, the configuration is symmetric—the ball and the hat appear identical from all sides. But the ball won't stay perched at the top for very long! To lower the system's energy, the ball will roll down the hill in one direction or another. It could roll in any direction, but it has to pick some direction: At that point, the symmetry becomes broken. Spontaneous symmetry breaking describes a system where the lowest energy state has less symmetry than the equations that describe that system.

Nature has many other examples. Another easy one to picture is a broom handle that is balanced, standing vertically on one end on a flat (circular) table. The equations that describe this system are completely symmetric with respect to rotations about the axis defined by the vertical broom, but when the broom falls over, it must fall in some direction and thus break this symmetry spontaneously. Likewise, any chunk of magnetized iron is an example of spontaneous symmetry breaking. When the iron is molten the spins of the individual atoms point in all directions and the equations describing their interactions have rotational symmetry, but as the iron cools it has a lowest energy state in which the spins are predominantly aligned in some direction, giving the iron a magnetic axis that breaks the rotational symmetry.

The symmetry that is broken in particle physics is the symmetry between the different particle types of the electroweak force—the photon, the W boson, and the Z boson. Experimentally, they look completely different. We see photons with our eyes, but it takes accelerators to detect W and Z bosons. Yet the fundamental equations describing these different particles (and the forces they mediate) are almost the same.

The difference is largely responsible for the nature of our universe. As go the particles, so go the forces that they mediate. Because of the symmetry breaking between the photons and the W and Z boson, electricity (mediated by the photon) is the basis of the modern world, and weak forces (mediated by the W and Z bosons) are mostly hidden inside individual atoms.

By discovering the Higgs particle at accelerators, or possibly something more complex, physicists hope to learn how nature broke the symmetry between the different particles and forces.

Model is mathematically inconsistent if the Higgs particle—or whatever replaces it—is too much heavier than the W and Z. Thus, combined with experimental measurements, the Higgs particle should weigh no more than around 300 GeV. It may be within reach of experiments at Fermilab's Tevatron, and it is certainly within reach of the Large Hadron Collider (LHC) currently being constructed at CERN.

Two more potentially important approaches to unifying the particle forces are "grand unification" and "supersymmetry." These ideas, which are explained in more detail below, are responsible for a good deal of the excitement about potential new discoveries at the Terascale.

Grand unification is the idea that all three of the Standard Model interactions (the weak, electromagnetic, and strong forces) are different aspects of a single larger set of interactions that has a larger, but spontaneously broken, symmetry. One powerful argument in favor of this idea is that the coupling strengths of the different interactions change with energy, and all appear to become roughly the same at a very high energy scale. Furthermore, the distinct types of particles observed in nature fit together beautifully in the larger symmetry patterns predicted by grand unification. Some signatures of grand unification may be accessible to experimental study at the Terascale, and others are best investigated by experiments that probe neutrino masses, the polarization of the cosmic microwave radiation, proton decay, and other rare or unusual phenomena.

Supersymmetry is a new type of symmetry that uses quantum variables to describe space and time. If supersymmetry is a symmetry of our world, space and time have new quantum dimensions as well as the familiar dimensions that we see in everyday life. Ordinary particles vibrating in the new quantum dimensions would then appear as new elementary particles, which could be detected using accelerators. Supersymmetry suggests that every known particle has an as-yet-undiscovered superpartner particle. If the symmetry were exact, the partners would have mass equal to that of the observed particles. This is not the case (or the superpartners would already have been observed), so this symmetry, too, must be broken.

Why do particle physicists think that supersymmetry is likely to be correct? The reason is that without it, it is very hard to understand how the scale of electroweak symmetry breaking (characterized by the W, Z, and Higgs boson masses) can be so small compared to the scale of possible unification, where the strengths of the strong, weak, and electromagnetic forces become equal. That is, above the scale of symmetry breaking between the electromagnetic and weak forces, one would expect the strengths of the forces to be equivalent, but this only happens at a much higher energy scale. Thus, supersymmetry makes it possible to understand why the W and Z have masses around 100 GeV. In addition, supersymmetry makes the unification of the three couplings occur more precisely. Of the superpartners predicted by supersymmetry, the lightest neutral superpartner particle, a neutralino, is thought to be an excellent candidate to account for some or all of the dark matter in the universe. Theoretical arguments strongly suggest that some of the new supersymmetric particles will be produced at the LHC. Supersymmetry is one of the most stimulating and challenging new ideas that physicists will be exploring in the Terascale regime.

There is one more important force in nature that is not usually regarded as a particle force, because its effects are unmeasurably small for individual elementary particles. This is gravity, which is the dominant force for stars, galaxies, and the universe as a whole but is so weak at the atomic level that it is not included in the Standard Model. Nevertheless, gravity is actually very similar to the other forces in that the mathematics of the Standard Model is stunningly similar to the math-

BOX 2-2
String Theory

An idea that may someday result in a full unification of all the forces appeared on the scene in the 1970s. Known as string theory, the idea in its most naïve form says that an elementary particle is not a point particle but a loop or a strand of vibrating string. Like a violin or piano string, one of these strings can vibrate with many different shapes or forms. In string theory the different forms of vibration of the string correspond to the various elementary particles—electrons, neutrinos, quarks, W particles, and so on. Unification of all forms of matter and of all the forces is achieved because the different matter particles and the carriers of the forces all arise from different forms of vibration of the same string.

Can string theory be tested? One testable idea associated with string theory is supersymmetry. Supersymmetry can exist without string theory, but string theories almost always have supersymmetry, and, indeed, the idea arose from early string theory work. Discovering the superpartner particles associated with supersymmetry at accelerators would help update relativity in the light of quantum mechanics and would give a major boost to string theory.

ematics used to describe gravity in Einstein's general theory of relativity. Thus, in contemporary physics, all the known forces are described in very similar ways.

Are these forces merely similar, or does this similarity point toward a truly unified theory that includes gravity as well as the particle forces? Within the usual theoretical framework, the differences lead to an impasse, and no combination of the two theories, the Standard Model and Einstein's general relativity, can be found. Understanding how to combine quantum mechanics and gravity is one of the goals of string theory (see Box 2-2). Combining quantum mechanics and gravity and finding ways to experimentally test these ideas are big challenges. Yet these challenges must be met to understand the development of the universe.

WHAT DO THE PROPERTIES OF PARTICLES REVEAL ABOUT THE NATURE AND ORIGIN OF MATTER AND THE PROPERTIES OF SPACE AND TIME?

Though particle physics focuses on the fundamental particles of the universe, it involves far more than just developing a taxonomy of esoteric phenomena studied in accelerator laboratories. An underlying quest of particle physics has been to understand how the detailed properties of particles and their interactions have influenced (and, in turn, been influenced by) the evolution of the cosmos.

At the beginning of the 20th century, the electron, which had just been discovered, was the only known particle that is today considered elementary. But the newly discovered phenomenon of atomic radioactivity gave physicists their first access to particles that, by the standards of the day, had high energies. (The energy of a particle emitted by a radioactive atom is about a million times greater than the energy of an electron that comes from a battery—and it is a million times smaller than the highest energy reached in modern particle accelerators.) With particle beams from naturally occurring radioactive sources, physicists made a host of major discoveries. The atomic nucleus, the proton, and the neutron all were discovered in this way, and the existence of the neutrino was inferred from studies of atomic radioactivity.

A new source of naturally occurring particles was discovered in 1912: Earth is constantly bombarded with cosmic rays from space. Besides giving physicists a fascinating new window from which to explore the universe, cosmic rays made possible fundamental discoveries about nature, mainly because cosmic rays have higher energies than do the particles emitted by radioactive atoms. The first antimatter particle, the positron (which is the antiparticle of the electron), was discovered in cosmic rays in 1932. Other important particles, including the muon, the pion, and the first strange particles, were discovered in cosmic rays in the 1940s and 1950s.

By then it was clear that many surprises lurked in the subatomic world. Beginning in the 1950s, man-made particle accelerators made it possible to achieve a combination of high energy and precision that could not be reached with naturally occurring particle sources. The first results brought chaos in the 1950s and the 1960s, as accelerators discovered literally hundreds of new kinds of particles that experience the strong force that holds the atomic nucleus together. All these particles are cousins of the familiar proton and neutron, which are the building blocks of atomic nuclei.

The Standard Model, which emerged in the early 1970s, brought some order to this chaos (see Box 2-3). According to the Standard Model, the multitude of particles arises by combining in many different ways a much smaller number of more fundamental entities called quarks. The strong force, which is mediated by particles known as gluons, binds the quarks together to form protons, neutrons, and other strongly interacting particles. Within atomic nuclei, the strong force arises as a consequence of the quarks and gluons in one neutron or proton interacting with those of another. The existence of quarks was confirmed in electron scattering experiments at the Stanford Linear Accelerator Center (SLAC) and in neutrino scattering experiments at CERN in the early 1970s. The gluon particle that binds together quarks was discovered at the Deutsches Elektronen-Synchrotron laboratory (DESY) in Germany in 1979.

Reinterpreting the multitude of particles produced at accelerators in terms of quarks and gluons gave a simpler explanation of how nature works. It also gave an entirely new foundation for thinking about unification of the forces of nature.

BOX 2-3
Particles in the Standard Model

The Standard Model contains six quarks with the names up (u), down (d), charm (c), strange (s), top (t), and bottom (b). The quarks are classified into three families or generations. The Standard Model also contains three electrically charged leptons—the electron (e), the muon (μ), and the tau (τ)—and three uncharged neutrinos (v_e, v_μ, and v_τ) (see Figure 2-3-1). The particles in each higher generation are identical to the particles in the previous generation except they have higher masses and quickly decay into other particles.

Understanding why there are three generations is an open question. Experimental evidence suggests there can only be three light neutrinos and so only three generations of particles. At the same time, quantum mechanics shows that three generations is the minimum number that can accommodate a mechanism known as CP violation, which allows matter and antimatter to behave slightly differently and which may have been critical in the formation and evolution of the universe.

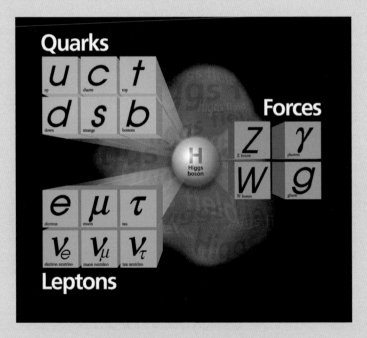

FIGURE 2-3-1 The particles of the Standard Model. Recent discoveries about neutrinos indicate that this picture of the Standard Model must be revised. The neutrino states (the v_e, v_μ, and v_τ) in this old version are now understood not to be particles with definite mass like their partners in each generation, the quarks and charged leptons. Instead, the v_e, v_μ, and v_τ are mixtures (combinations) of definite-mass neutrinos, which leads to the phenomenon of neutrino oscillations, which has recently been established experimentally. This discovery requires the first significant revision of the Standard Model in decades. Courtesy of Fermilab Visual Media Services.

Quarks obey equations similar to the equations obeyed by electrons, and gluons obey equations similar to the equations obeyed by photons, or light waves. The analogy was further improved when the CERN accelerator in 1983 discovered the W and Z particles, which are responsible for the weak atomic force and obey the same sort of equations as gluons or photons. At DESY in the 1990s, the properties of the strong force and the numbers and energy distributions of quarks and gluons in high-speed protons were measured with great precision; these results have been important inputs into the expectations for LHC physics. Again, new discoveries made at high energies showed that at a fundamental level the different forces are all very similar, giving physicists a new foundation for seeking to unify the laws of nature.

The Standard Model further reduces the observed complexity of particles by organizing quarks and leptons (the most familiar of which is the electron) into three "generations." The first generation contains the particles making up ordinary atoms—the up and down quarks and the electron, along with a more elusive entity called the electron neutrino. Such neutrinos are created in the radioactive decays of certain types of nuclei. Neutrinos interact so weakly with matter that when they were first hypothesized in the 1930s, physicists thought they would be undetectable. The invention of nuclear reactors changed the situation by making available intense sources of electron antineutrinos, leading to the detection of the neutrino in 1955.

One generation of particles would suffice to construct ordinary matter. Oddly, nature repeats itself with two more generations of particles. These additional particles, which are short-lived, are usually only produced in high-energy collisions and detected by their decay remnants. While they are subject to precisely the same forces as the first-generation particles, they decay so quickly that they are harder to study. But in the early universe they appear to have been just as important as the first-generation particles. Physicists do not yet understand why particle generations exist, much less why there are three of them.

It is, however, believed that the number of generations is precisely three. The best indication of this comes from studies of the Z particle, which carries the weak force. All types of neutrinos can be produced when the Z particle decays, provided that they are less massive than half the Z mass. The pattern of Z production and decay shows that it decays into only three types of neutrinos, so a fourth neutrino type can exist only if it is very heavy. The amount of helium produced in the early universe is also sensitive to the number of neutrino types, and measurements of this abundance are consistent with the existence of just three types of light neutrinos. Since all known neutrino types are very light, this tells us that there is no fourth generation of particles that follows the same pattern as the first three with a very light neutrino.

First-generation particles are present all around us in ordinary matter. But how do we know about the other two generations? The discovery of the second generation began in the 1930s and 1940s, when the muon and mesons, which consist of a quark and an antiquark, were discovered in cosmic rays. When these high-energy particles from space strike the atmosphere, the collisions are energetic enough to produce many mesons containing the second-generation strange quark. The mesons then decay, many of them by a weak interaction process that produces a muon and a neutrino.

In the 1950s, particle accelerators with enough energy to create second-generation particles were built for studying the behavior of particles in controlled experiments. By 1962, using high-energy neutrino beams created at accelerators, the second-generation neutrino was discovered; an experiment at Brookhaven National Laboratory demonstrated that the neutrinos created along with muons in meson decays are distinct from the first-generation neutrinos created in decays of radioactive atoms. The discovery of the second generation was completed when evidence for the charm quark was found at particle accelerators, beginning with the discovery of the J/ψ particle (which consists of a charm quark and an anticharm quark) in November 1974 at SLAC and Brookhaven.

Experimental discovery of the third generation began when the tau lepton was discovered in 1975 at SLAC, after which particles containing the bottom quark were discovered in 1977 at Fermilab and at Cornell. Once the tau lepton and bottom quark were observed, the search began for the third-generation top quark. But what would it weigh? All that was known was that the top quark would have to be heavier than the bottom quark, or it would have been found at the energy levels already explored. The bottom quark weighs about 5 GeV, or about five times the mass of the proton (which contains three of the much lighter quarks).

By the early 1990s, experiments provided an indirect estimate of the mass of the top quark. Even if a particle is not produced in a given reaction, it can influence that reaction through quantum effects. According to quantum mechanics, particles and their antiparticles can wink in and out of existence for unobservably short times, thereby producing small but measurable effects on particle interactions. By the early 1990s, the data on the properties of Z bosons were precise enough to be sensitive to quantum effects due to top quarks. This led to an estimate that the top quark's mass was 150 to 200 GeV. For a mass outside this range, the measurements could not fit with Standard Model predictions.

This mass range was just barely in reach of the Tevatron, and in 1995 the top quark was discovered at Fermilab, with a mass since measured to be 174 GeV. The initial discovery was based on just a few dozen events, in which a top quark and antiquark were produced and decayed to other particles, including bottom quarks and leptons, in a characteristic and expected pattern.

Completion of the third generation required confirmation that the third generation has its own neutrino type. Thus, the neutrino produced in association with a tau particle should make only tau particles when it interacts with a W particle. This confirmation was achieved at Fermilab in 2000. With the tau neutrino observation, three of the four particles of the Standard Model's third generation had been discovered at Fermilab.

Observing neutrino effects is hard, but an even bigger challenge for particle physics has been to detect and measure the masses of the neutrinos. Those masses still have not been precisely determined, yet they are suspected to be very important clues about particle unification. There are several approaches to detecting neutrino masses, the most sensitive of which depends on the fact that there are multiple types of neutrinos. If neutrinos have mass, a quantum mechanical effect known as "neutrino oscillations" can come into play. As a neutrino of one type travels through space, it can spontaneously convert to another type. For example, a muon neutrino can convert spontaneously to a tau neutrino or to an electron neutrino. Later it can revert to being a muon neutrino, which is why neutrino types are said to oscillate. The probability for oscillation depends on the differences in the masses between the neutrinos, and it takes very large distances for these changes to occur with a high probability.

Neutrinos created in the sun travel 93 million miles before they reach Earth, which makes them likely candidates to undergo oscillations. Beginning with pioneering measurements made almost 40 years ago in the Homestake Gold Mine in South Dakota, every measurement of the number of electron neutrinos reaching Earth from the sun has given an unexpectedly small result. Subsequent observations, notably in laboratories in Japan and Canada, have found similar anomalies in the properties of neutrinos created in Earth's atmosphere by cosmic rays, neutrinos from nuclear reactors, and neutrinos produced in accelerators. All these observations are understood today in terms of neutrino masses and oscillations.

When the second generation first emerged—with the discovery of the muon in cosmic rays—it fell from the sky to everyone's surprise. I.I. Rabi famously asked: "Who ordered that?" By contrast, the existence of a third generation was suggested in advance as a possible explanation of what is called CP violation (see below).

One of the surprising predictions from combining quantum mechanics with special relativity is the existence of antimatter. Antimatter was first discovered in cosmic rays as antielectrons (positrons). The antiproton was first created artificially at one of the early high-energy accelerators, the Lawrence Berkeley National Laboratory Bevatron. For every type of particle, there is a corresponding antiparticle with the same mass and spin but with opposite electric charge. When particle and antiparticle meet, they can annihilate into radiation. The laws of physics for matter and antimatter are very similar, but in the universe today there is lots of matter and very little antimatter. The reason for this is a mystery.

In 1964 it was discovered at Brookhaven that matter and antimatter behave slightly differently. In this experiment, scientists prepared a beam of kaon particles such that it was about half matter and half antimatter. By carefully studying the particles, they observed that the matter particles behaved differently than the antimatter ones (see Box 2-4). This discovery was a great surprise, not only because it violated the presumed equivalence of matter and antimatter but also because it

---

BOX 2-4
Sakharov, Antimatter, and Proton Decay

Andrei Sakharov (1921-1989) is best known to the general public as the architect of the Soviet nuclear bomb who later became a fearless advocate of human rights and peace. The Nobel peace prize committee called him "a spokesman for the conscience of mankind." Many credit him with helping to end the cold war.

Sakharov also conceived the idea of building a toroidal magnetic coil (tokamak) to generate fusion energy. This is a key concept behind the current international fusion energy collaboration known as ITER.

But particle physicists also know Sakharov for a daring cosmological proposal he made in 1967. Sakharov wanted to explain why the universe seems to be filled with matter, while antimatter is nowhere to be seen—except when it is produced by cosmic rays or radioactive atoms. Since matter and antimatter seem to have equivalent properties, why is the universe filled with one and not the other?

Sakharov's concept was that the very early universe was filled with a huge density of matter and antimatter at a vast temperature. The temperature and the density that Sakharov assumed are far beyond anything that exists in the current universe, even at the center of stars or in particle accelerators.

Then, Sakharov said, as the universe expanded and cooled, almost all of the matter and antimatter annihilated and disappeared. But a slight asymmetry developed, and as the antimatter annihilated, a tiny bit of matter remained. From that small remnant of the cosmos's origin, according to Sakhavov, stars, planets, and people ultimately formed.

For this to work, Sakharov showed, two very subtle particle physics effects would be needed. The first is a tiny asymmetry between the behavior of matter and antimatter that is known as CP violation. This had been discovered in 1964 at Brookhaven National Laboratory in careful studies of the decays of elementary particles known as kaons, or K particles. That discovery provided an important clue for Sakharov's work. Intensive studies of CP violation continue to the present day, notably at the B factories at SLAC in California and at KEK in Japan.

Sakharov boldly predicted a second subtle effect that is needed for his approach to cosmology to work: The proton cannot live forever. It must decay. Thus, all ordinary atoms (which contain protons in their nuclei) must ultimately decay.

Intensive experimental searches for proton decay have not yet been successful. Yet, since Sakharov's work, new reasons have emerged to suspect that the proton does decay. The search goes on.

suggested a connection between the microphysics of elementary particles and the macrophysical question of the amount of antimatter in the universe. This small but fundamental asymmetry in physical laws between matter and antimatter is the above-mentioned CP (or charge parity) symmetry violation. Since then, important experiments at Fermilab in 1999 studied the kaon system further and confirmed the presence of CP violation not only in the behavior of the kaons but also in their decays.

The early universe was filled with matter and antimatter, and modern particle and cosmology theory strongly suggest that they both were equally represented. As the universe cooled, matter and antimatter annihilated each other. If the laws of nature had had perfect symmetry between matter and antimatter, the cooling universe would have maintained equal amounts of matter and antimatter, which would have been capable of completely annihilating into photons. By the time "ordinary" temperatures were reached (in this context, a million degrees is low enough), the matter and antimatter would all have disappeared, leaving only photons and dark matter. The result would have been a very dull universe.

Instead, the early universe seems to have produced a slight excess of matter over antimatter. After the bulk of the matter and antimatter annihilated each other, only this excess remained. Today the universe contains more than a billion photons for every proton, neutron, and electron. In the overall universe, therefore, the leftover matter is just a trace, but it has condensed into dense regions to form galaxies, stars, and planets.

In the Standard Model, CP violation cannot occur in a two-generation world; it requires a third generation. With the third generation included, the Standard Model leads to an elegant theory of CP violation. To test it effectively requires experiments with third-generation particles, because CP-violating effects are so tiny for the first two generations.

A particle called the B meson (a particle containing one b quark and one lighter quark or antiquark) is the right one for the job. Careful study at the Cornell Electron Storage Ring (CESR) and at the DORIS storage ring at DESY in Hamburg showed that the B meson can change into an anti-B meson (and back again) and that the bottom quark undergoes rare weak decays to an up quark, both of which are key if CP violation is to be observable. Furthermore, experiments at particle accelerators showed that the B meson survives a trillionth of a second before decaying, which is surprisingly long for such a massive particle. This is long enough for CP-violating effects to take place—and for them to be observed.

The critical advance in this area was the construction of "B factories" at SLAC and at KEK in Japan. A B factory is an electron-positron collider designed to create a very large number of B mesons. In fact, these accelerators have the greatest luminosity, measured by the rate of particle-antiparticle collisions, of any accelerator ever built. This high rate is needed because the study of CP violation depends on recording many very rare particle decays.

The B factories incorporate novel techniques to make these experiments feasible. The beams are asymmetric (one of the colliding beams has more energy than the other), so that the resulting B particles are produced with high velocity. This makes it possible to measure the tiny times (corresponding to flight-length distances of a few hundred microns in the detector) involved in B particle decays. The B factories have made many important new measurements of CP violation. These measurements fit together exactly as expected by the Standard Model, providing a unique precision test of its predictions.[2]

However, although measurements of CP violation at the B factories matched the Standard Model, they cannot account for the asymmetry in the amounts of matter and antimatter in the universe. That is, cosmological observations about the relative abundances of matter and antimatter in the universe are not explained by Standard Model physics of the early universe.

Extremely precise measurements of parameters provide extremely sensitive tests of particle theory: Thus, extra precision has important dividends. There are several avenues to achieving greater precision. In some cases, greater precision is possible by collecting more data, which may require more intense particle beams, more sensitive equipment, or other technical advances. In other instances, one must develop new techniques of measurement or a capability for performing entirely new types of experiments.

Measurements of weak interactions over the past decade provide a good example of the usefulness of large data samples. Since electrons and positrons are relatively simple, well-understood particles, the greatest precision in testing detailed predictions often has come from experiments using them. At the beginning of the 1990s, the energy available in electron-positron collisions reached the mass of the Z particle. This energy became available in electron-positron collisions at the Stanford Linear Collider (SLC) at SLAC and at the Large Electron and Positron (LEP) collider at CERN with higher luminosity. After a few years of LEP data, measurements of Z particles became available based on millions of events rather than thousands.

---

[2]The two experiments at Fermilab's Tevatron, CDF and D0, have recently announced the first measurements of the mixing frequency for a special type of B particle, the B_s. These observations of the properties of this subatomic particle suggest that it oscillates between matter and antimatter in one of nature's fastest rapid-fire processes—many trillions of times per second. Please see <http://www.fnal.gov/pub/presspass/press_releases/CDF_04-11-06.html> and <http://www.fnal.gov/pub/presspass/press_releases/DZeroB_s.html> for more information.

Another improvement in precision measurements of the weak interactions came after 1995, when the energy of LEP was doubled by adding accelerating cavities to the machine. This made it possible to produce the W particle in pairs and to measure the mass and properties of the W more precisely. Along with measurements of the W and top quark masses at Fermilab, the production of W particles in pairs led to indirect estimates of the mass of the Higgs particle in the Standard Model, due to its quantum effects on these quantities. For all the pieces to fit together with a single set of Standard Model parameters, a Higgs particle must exist with a mass below about 300 GeV. If experiments at the LHC do not discover a Higgs particle within the expected range, the mechanism that produces particle masses must be more complicated than the hypothesis incorporating a single Higgs boson.

As the examples in the previous paragraphs demonstrate, precise measurements in particle physics do not always require the highest possible energies to probe for new physics effects (see Box 2-5). New particles or processes that can only be directly observed at very high energies can cause effects at lower energies. Such effects could change the decay properties of lighter particles containing strange, charm, or bottom quarks from the predictions of the Standard Model. As long as all the various measurements taken together fit Standard Model predictions, they also provide lower limits on the masses, or combinations of masses and couplings, of any particles that may exist at very high energies, because any such particles would contribute to all decays via these quantum effects.

Thus, precision measurements are windows onto energies far above those that can be created in the laboratory. By comparing the experimental measurements with predictions from the Standard Model, particle physicists look for tiny deviations from Standard Model predictions. Any such deviations can be interpreted as signals for particles not in the Standard Model that exist at a higher energy scale than is possible to produce directly at an accelerator. These deviations also could be interpreted as signals for new physical structures of the universe such as dimensions in addition to the three we observe with our eyes.

Experiments using beams of muons or kaons (which contain strange quarks) and experiments observing the decay of D mesons (which contain charm quarks) have ruled out some departures from the Standard Model as small as one part in a trillion, eliminating many models with unseen particles at masses up to the Terascale. Similarly, CESR and the B factories at SLAC and KEK have used decays of bottom quarks to rule out other hypotheses. Important limits on transitions from one type of charged lepton to another that do not include the expected types of neutrino partners have been established at HERA in Germany, the Los Alamos National Laboratory, and the Paul Scherrer Institute in Switzerland.

In a similar fashion, neutrino masses can be a window onto unknown physics

---

BOX 2-5
Magnetic Moments of the Leptons: A Precision Measurement

Historically, one of the first very precise measurements in particle physics was the measurement of the magnetic moment of the electron in 1950, originally with a precision of about one part in a thousand. The electron has a tiny spin, like a quantum gyroscope, and also behaves as a tiny magnet, giving it a magnetic moment. Over the years, measurements of the electron's magnetic moment have improved, as have theoretical calculations. Measurement of the magnetic moment is important because of the sensitivity to phenomena that are not yet understood; that is, the rare opportunity offered in measuring the magnetic moment is not only that experimentalists can extract a very precise value but that theorists can make a very precise prediction using the tools of the Standard Model. By comparing these two results (the observed value and the predicted value) to high precision, particle physicists have constructed a very sensitive test of the accuracy of the Standard Model. (In general, just because a theorist can make a prediction doesn't mean an experimenter can prove whether that prediction is right or wrong!)

The latest measurement of the electron magnetic moment, reported in 2004, has an accuracy of better than one part in a trillion, which is perhaps the greatest precision with which any physical quantity has ever been measured. This measurement does not require an accelerator. It is made with a single electron stored in a tabletop device, in which the electron can be manipulated and studied with great precision for a long time. Because this quantity also can be calculated in the Standard Model to a high level of precision, the data and theoretical calculations together provide one of the most sensitive tests available today of the form of the new physics and the precise constraints on it.

The muon (the second-generation cousin of the electron) also has a magnetic moment. However, the techniques for measuring it are rather different, since muons are short-lived and can be produced only at accelerators. The most recent measurement of the muon magnetic moment, reported at Brookhaven in 2004, has a precision of about one part in ten billion, which also places this measure among the most precise in nature.

Comparisons between the experimentally measured magnetic moments of the electron and muon and the precise predictions of the Standard Model place important restrictions on the allowed masses of the new particles predicted in some extensions of the Standard Model.

---

occurring at high energy scales. Nonzero neutrino masses can be accommodated through models that contain new and as-yet-undiscovered particles. Detailed studies of the patterns of neutrino masses can give insight into physics at the high energy scales where these new particles are presumed to exist. Such particles are a necessary component of some models of unified forces and are predicted to exist at an energy scale far past the range of any foreseeable accelerator. Neutrino masses also open the door to CP violation in the neutrino world, similar to that already seen in the quark sector. This understanding of CP violation in the neutrino sector may lead to new explanations of how matter came to dominate antimatter in the early universe.

WHAT ARE DARK ENERGY AND DARK MATTER AND HOW HAS QUANTUM MECHANICS INFLUENCED THE STRUCTURE OF THE UNIVERSE?

Astronomers looking at the night sky used to assume that what they saw was pretty much what there was. Then, in 1933, astronomers studied the motion of galaxies and found that they were moving much faster than could be explained by the known gravitational forces due to other nearby galaxies. This was the beginning of the dark matter problem. To account for the unexpectedly rapid motion of the galaxies—and, as later became clear, the rapid motion of individual stars making up galaxies—one must assume that galaxies are surrounded by clouds of dark matter. In recent times, scientists have found more and more ways to observe the gravitational effects of dark matter but have not yet learned what the dark matter is. All that is known for sure about the dark matter cloud surrounding a galaxy is that, typically, it is considerably larger and heavier than the visible part of the galaxy. In fact, according to the most recent measurements by NASA's Wilkinson Microwave Anisotropy Probe (WMAP) satellite, dark matter accounts for about six times as much of the universe as the ordinary matter that can be seen.

Different theories of dark matter have led to different strategies for detecting it, none of which have been successful so far. If dark matter is a cloud of elementary particles, it may be detectable in sensitive particle detectors placed deep underground for shielding from ordinary cosmic rays. Calculations show that a cloud of Terascale particles would have just about the right properties to agree with what is known about dark matter. Underground laboratories are approaching the sensitivity at which such a cloud could be detected, so there is a chance of uncovering the nature and properties of dark matter in the near future.

Dark matter is only one of the surprising discoveries made by astronomers about the content of the universe. Since the discovery in the 1920s that the universe is expanding, astronomers and physicists have assumed that the expansion is slowing because of the gravitational attraction between galaxies. Numerous attempts were made to measure this presumed deceleration of the cosmic expansion, but the attempts were frustrated by the difficulty of estimating precise distances to remote galaxies.

Then, in the 1990s, measurements of large-scale structure in the universe, including clusters and superclusters of galaxies, and of the radiation that permeates the universe suggested that most of the energy in the universe consists of dark energy, a smoothly distributed, all-pervasive form of energy that causes the expansion of the universe to accelerate. Supernovae in distant galaxies were also used to gauge cosmic distances and provided direct evidence that the expansion of the universe is speeding up.

The dark energy responsible for this accelerated expansion of the universe might be interpreted theoretically in terms of what Einstein called the cosmological constant. It is not yet clear whether this interpretation is correct or whether some more elaborate theory of dark energy is needed. In any event, the acceleration of the cosmic expansion calls for a fundamental modification of existing ideas about nature. Calculations of the amount of dark energy in the Standard Model using the most reasonable assumptions differ from the experimental result by at least 60 orders of magnitude! Obviously, the current understanding of the situation is incomplete.

This problem is closely related to the effort to unify the Standard Model and general relativity. Indeed, the problem of dark energy combines considerations of quantum mechanics, which contributes to the vacuum energy via quantum fluctuations, with Einstein's theory of gravity, without which the energy of the vacuum would be unobservable. No formalism has yet been devised that combines the theory of gravity and quantum mechanics in a satisfactory way.

The overwhelming scientific interest in dark matter and dark energy is driven by the fact that these seemingly exotic substances were discovered because of their very real effects on the structure and evolution of the universe.

Another challenging idea about cosmology is the idea of the inflationary universe, which is closely linked to particle physics. According to this hypothesis, the vast and nearly homogeneous universe that we see today originated in a period soon after the big bang, when the universe underwent a period of accelerated expansion that was 100 orders of magnitude faster than the acceleration due to dark energy. The cause of this rapid expansion is thought to be a field dubbed the "inflaton," which dominated the universe for a brief instant after the big bang and then disintegrated into the matter and radiation observed today. During that brief period, the inflaton stretched the universe by a factor of 10^{100} or more, making it smooth and flat. However, the quantum process associated with the disappearance of the inflaton field caused the distribution of the remaining energy and radiation to be slightly nonuniform after the inflation was complete.

Surprisingly, it has proved possible to test the inflationary hypothesis using the fact that space is filled with a diffuse radiation called the cosmic microwave background (CMB), which is interpreted as radiation that was created at the beginning of the universe. The CMB has a temperature of about 2.7 K. This is the temperature that one would measure if one placed a thermometer in outer space far from any star. The CMB is highly isotropic, which means that the temperature appears to be nearly the same no matter in which direction one looks (see Box 2-6).

However, in the 1970s, researchers realized that to account for the formation of clusters of galaxies, the CMB must be slightly anisotropic, with slightly different temperatures in different regions of space. When the idea of inflation was intro-

BOX 2-6
The Cosmic Microwave Background: Footprints of the Early Universe

The 21st century will be the first time in history when humans view the universe with high precision all the way out to the cosmic horizon. The light traveling from the most distant reaches of space will provide detailed information about the universe in its early stages, when the temperature and density of the universe exceeded what can be achieved at the highest energy accelerators imaginable, including the LHC and the ILC. For this reason, cosmology and elementary particle physics have become intimately intertwined, providing information that simultaneously improves understanding of both the smallest and largest entities in the cosmos.

Breakthroughs in cosmology have been made possible by a confluence of new, highly advanced technologies. For example, the first highly precise microwave, infrared, and x-ray surveys of the distant universe have been completed; the three-dimensional structure of the nearby universe has been mapped out by the first red shift surveys; and views of the first stars and galaxies have been captured by the Hubble Space Telescope and by giant segmented-mirror telescopes on the ground.

The snapshot of the infant universe taken by the WMAP satellite—sure to be one of the icons of 21st century science—is emblematic of this generation of powerful new probes (see Figure 2-6-1). In early 2006, a new, more detailed picture of the infant universe was released. Colors indicate "warmer" (red) and "cooler" (blue) spots. The white bars show the polarization direction of the oldest light. This seemingly formless pattern is chock-full of valuable information. First, it shows in detail the distribution of energy in the universe more than 13 billion years ago, when the first atoms formed. In this pattern can be identified the regions that later gave birth to galaxies like our own Milky Way (red and yellow) or that grew into giant, nearly vacuous voids (blue). Second, by studying how the number of spots and energy concentration vary with the spot size, cosmologists can derive a precise measure of the composition of the universe, providing the best evidence that the universe contains 4 percent ordinary matter, 20 percent dark matter, and more than 75 percent dark energy. Perhaps most exciting is the information an improved map of the cosmic background radiation and forthcoming measurements of its polarization will provide about the events that created the splotches in the first place. The measurements may prove that inflation accounts for the structure of the universe, as most cosmologists believe, and provide insights about the ultra-high-energy physics effects that caused inflation.

FIGURE 2-6-1 Infant universe as seen by WMAP. Courtesy of the NASA/WMAP Science Team.

duced a decade later, it was realized that the early periods of inflation solved the problem of generating the density lumps need to explain the splotches in the CMB. In 1992, a dedicated satellite experiment, the Cosmic Background Explorer (COBE), was launched, and the predicted anisotropy of the background radiation was indeed detected. The temperature differences among the splotches were found to be only a few hundred-thousandths of a degree. A much more precise measurement of the temperature variations was made by WMAP in 2003 and in 2006. The temperature pattern found thus far is in excellent accord with the inflationary prediction, although the inflationary picture is not the only hypothesis consistent with the data.

To confirm the inflationary universe hypothesis (or an alternative) requires improved instruments that can make more precise measurements of how the CMB radiation varies with position in the sky. Among the most important tests will be measurements of the polarization pattern of the CMB radiation. The polarization of an electromagnetic wave is the direction along which its electric field oscillates. When the CMB radiation scatters from the sea of electrons and begins to stream toward us, it becomes polarized by an amount that depends on the cosmological model. This polarization was observed for the first time in 2002, but not yet with the sensitivity needed to definitively test the inflationary theory of the universe.

ROLES OF ACCELERATOR- AND NON-ACCELERATOR-BASED EXPERIMENTS

This recent history of particle physics underscores the interplay between experiments involving accelerators and those that do not involve accelerators. For instance, nonaccelerator experiments have helped drive the scientific frontiers of particle physics and have brought the field into closer contact with nuclear physics, cosmology, and astrophysics. Historically, many important discoveries first came from nonaccelerator experiments, in some cases simply because appropriate accelerators did not exist at the time. In fact, there is an impressive list of particle physics discoveries that did not involve accelerators. To name just a few:

- Discovery of the neutron,
- First evidence for the neutrino,
- Detection of antimatter (discovery of the positron),
- Discovery of parity violation,
- Detailed exploration of the weak interaction,
- Discovery of muons,
- Discovery of pions,
- Discovery of V particles (later called kaons),

- Direct detection of neutrinos, and
- Recently, discovery of neutrino mass and mixing.

Accelerator-based experiments, on the other hand, have also led to important accomplishments:

- Discovery of the electron.
- Discovery of the composite nature of the proton.
- The era of particle "zoology" in the mid-1900s, when 100 or so particles and resonances were found, which in turn called for a simpler framework. The ultimate solution was the proposal of the quark model.
- Discovery of the antiproton and antineutron. This discovery validated and solidified the Dirac theory of antiparticles. Even though the positron, found in cosmic rays, was already known to exist, physicists were not sure about every particle having an antiparticle.
- Discovery of the K_L meson. Although the kaon was discovered in cosmic rays, its rich physical properties were elucidated in accelerators. Resolution of the puzzle of a particle with two lifetimes led to the discovery of CP violation. CP violation was key to the attempt to understand matter-anti-matter asymmetry in the universe.
- Discovery of the second-generation neutrino.
- The 1974 discovery at SLAC and Brookhaven National Laboratory of the charm quark.
- Discovery of jets in electron and positron collisions.
- Discovery of the gluon at DESY, and confirmation of the existence of a particle carrying the strong force.
- Discovery at CERN of the W and Z, the carriers of the weak force, and confirmation of the gauge theory of weak interaction.
- Discovery of the tau lepton at SLAC, proving that there must be at least three families of leptons.
- Discovery of bottom quark particles at Fermilab and Cornell, demonstrating that there are three families of quarks in parallel with three families of leptons.
- Precision tests of the Standard Model using measurements at LEP, SLAC, Fermilab, DESY, CESR, and elsewhere (for example, measurements of particle masses, quantum numbers, and quark couplings and mixing rates).
- Measurement at LEP of light neutrino families that couple to the Z, showing that there can only be three families.
- Discovery of the top quark at Fermilab, completing the Standard Model.

- More recently, confirmation of neutrino oscillations and mixing through accelerator-based experiments at KEK in Japan and at Fermilab.

Without the contributions from accelerator experiments, modern particle physics would be far less advanced than it is today. There is no question that accelerators have been essential in particle physics and there is a clear role for them in uncovering the secrets of the Terascale. Indeed, much of the drama surrounding the Terascale comes from the expectation that accelerators will finally expose and then directly investigate the cracks in the Standard Model.

A remarkable aspect of particle physics today is that answers to many of the questions described in this chapter are within the reach of tools that can be built with currently available technologies. The next chapter explores the tools that will be available in the next decade to address the grand questions of particle physics, including those that will enable exploration of the Terascale.

3

The Experimental Opportunities

As described in Chapter 2, recent discoveries in particle physics have led to the key scientific challenges that now define the frontiers of research in the field. This chapter looks at experiments that could be done in the coming decade to address these exciting research challenges. Some of the facilities needed to carry out the next generation of experiments are now being built, such as the Large Hadron Collider (LHC) at the European Organization for Nuclear Research (CERN), new experimental facilities at the Japan Proton Accelerator Research Complex (J-PARC), experimental devices designed to measure cosmic microwave background (CMB) radiation, detectors for high-energy particles from cosmic sources, and instruments to detect gravity waves. Other key experimental facilities—such as the proposed International Linear Collider (ILC); enhanced neutrino studies at accelerators, at reactors, and in large underground laboratories; proton decay experiments; and new space-based experiments—are the subject of planning and ongoing research and development.

This chapter divides potential experiments into three categories: those using high-energy beams, those using high-intensity beams, and those using particle sources provided by nature. As is the case throughout particle physics, different experiments can address the same questions from different perspectives, revealing the rich interconnections within the field and between particle physics and other fields. The chapter concludes by outlining the increasing importance of international collaboration in particle physics—collaboration that best meets the needs of science and represents the most responsible public policy.

As the preceding chapter demonstrated, particle physics has entered a special time. The most exciting scientific questions that need to be addressed are clear. The next cohort of experiments needed to address many of those questions are about to begin or are on the scientific horizon. Expert groups of scientists, engineers, and advanced students are available and eager to move this segment of the scientific frontier forward. A goal that has occupied science for centuries—gaining a fuller and deeper understanding of the origins and nature of matter, energy, space, and time—is ready for what may be a revolutionary leap forward.

HIGH-ENERGY BEAMS: DIRECT EXPLORATION OF THE TERASCALE

Discoveries at the Terascale

With experimental study of the Terascale about to begin, physicists are finally gaining the tools needed to address questions that have been asked for decades:

- Why do the weak interactions look so different from electromagnetism, given that the fundamental equations are so similar?
- Where do particle masses come from? Does the Standard Model describe them correctly, or do the particle masses come from some more exotic mechanism?
- Are the forces of nature unified at some high energy scale? With the elementary particles known today, unification does not quite work, but it fails in a way that suggests the missing pieces will be found at the Terascale.
- Do space and time have additional dimensions? Do they have new quantum dimensions?
- What is the dark matter of the universe? Can it be produced in the laboratory?

The next generation of experiments will answer at least some of these questions.

Tools for Exploring the Terascale

Particle accelerators recreate the particles and phenomena of the very early universe. When particles collide in accelerators, new particles not readily found in nature can be produced and new interactions can be observed. These new particles and interactions were prominent in the early universe but disappeared as it cooled, leaving only scattered clues about their continuing influence. Understanding the properties of these particles, however, is essential to building a full understanding of the natural world and its evolution. Accelerator experiments are the sole places where these particles and interactions can be studied in a controlled fashion. Other facilities provide crucial information, but high-energy particle accelerators remain the most important single tool available for addressing the scientific challenges facing particle physics.

The Tevatron collider at Fermi National Accelerator Laboratory (Fermilab) outside Chicago is currently the highest energy accelerator in the world, and it will remain so for another year or two. The Tevatron collides beams of protons and antiprotons with a total energy of about 2 trillion electron volts (TeV). The luminosity, or intensity, of the particle beams at the Tevatron has steadily increased in the last few years, and continued increases are essential to the success of the Tevatron physics program. Precision measurements and discoveries at the Tevatron have helped to pave the way toward exploration of the Terascale at the LHC; measurement of the W boson mass and the discovery and measurement of top quark properties now help point the way toward the possible discovery of the Higgs particle and even supersymmetry at the Terascale.

The program at the Tevatron has two main thrusts: searches for new particles and precise measurements of particle properties. In the latter category, for example, the Tevatron continues to improve knowledge of heavy particles such as the top quark, which was discovered at the Tevatron and whose large mass still places it out of reach of other facilities.

In 2007 the LHC at CERN is scheduled to begin accelerating beams of protons to a total energy of 14 TeV, thus exceeding the energy available at Fermilab by a factor of 7. In historical terms, this is a large jump in energy, which is made all the more exciting because so many clues point to the importance of the Terascale (see Box 3-1). With its initial luminosity, the LHC has wide potential for new discoveries. The prospects are so varied as to defy brief summary, but they include possible new elementary particle forces, the first evidence for supersymmetric particles, the discovery of a Higgs particle, and much more. The LHC's discovery capabilities will grow further when it achieves its full luminosity after a few years of operation.

What is the next step beyond the LHC? The advance of science proceeds on many fronts and requires many different kinds of tools. If one kind of tool were the best for all purposes, that tool would be built and then made bigger or better. But the world does not give up all of its secrets that way.

In particle physics the obvious needs are for higher energy, more accurate measurements, and the ability to detect new, rare, or elusive processes. Each of these frontiers is best advanced with a different kind of instrument.

To make an analogy, in astronomy the largest Earth-based telescopes are capable of detecting the dimmest objects; the Hubble Space Telescope (HST) has a smaller mirror but is able to produce the sharpest pictures; and numerous other

BOX 3-1
Particle Detectors

In particle physics, experimentation studies collisions of particles that have been accelerated to very high energies. The collisions convert energy to mass, producing new particles or new phenomena associated with fundamental particle interactions through Einstein's famous equation, $E = mc^2$. Particle physics facilities can be thought of as enormous microscopes that are powerful enough to probe physical processes at extremely small distance scales. In modern particle physics experiments, different types of detector systems surround the collision point. The detectors measure the properties of the passing particles.

The LHC, which is scheduled to begin operation in 2007, will produce proton beams seven times more powerful than those at Fermilab. The LHC beams also will reach much greater levels of intensity. In fact, experiments at the LHC will witness something like 1 billion collisions per second. Only 100 collisions per second, at 1 megabyte of data per collision, can be recorded for later analysis. It is a major challenge to design and build the high-speed, radiation-hardened custom electronics that provide the pattern recognition necessary to select potentially interesting collisions.

In a colliding-beam experiment, the particles travel out in all directions from the collision point, so the detector is usually as tightly closed as possible (see Figure 3-1-1). Following each

FIGURE 3-1-1 An artist's illustration of a particle collision event. Courtesy of the ATLAS experiment.

continued

BOX 3-1 Continued

collision, called an event, computers record the data. Each particle type has its own signature in the detector, but the detailed analysis of an event can be very complicated and can sometimes take years and a great deal of scientific creativity and judgment to decipher correctly. The results of these analyses generate the key scientific discoveries.

There are two multipurpose experiments at the LHC, the Atoroidal LHC Apparatus (ATLAS) and the Compact Muon Solenoid (CMS). The ATLAS experiment, the larger of the two, is about the size of a five-story building. ATLAS and CMS are the largest collaborative efforts ever attempted in the physical sciences. For example, at present ATLAS has more than 1,800 physicists (including 400 students) participating in the experiment from more than 150 universities and laboratories in 34 countries.

The two experiments are similar in concept but different in detail. ATLAS and CMS both have charged-particle tracking to determine particle momentum; calorimetry to measure the energy of electrons, photons, and quark jets; and the ability to identify muons. ATLAS detects muons with a gigantic toroid assembly. CMS detects electrons and photons with its crystal calorimeter. Both experiments can detect short-lived particles with silicon pixel vertex detectors. ATLAS and CMS are poised to make discoveries when the accelerator delivers its first collisions (see Figure 3-1-2).

Some interesting facts about CMS are as follows (ATLAS has its own set of fascinating facts):

- The total mass of CMS is approximately 12,500 tons—double that of ATLAS (even though ATLAS is about eight times the volume of CMS).
- The CMS silicon tracker comprises approximately 250 square meters of silicon detectors—about the area of a 25-meter-long swimming pool. The silicon pixel detector comprises more than 23 million detector elements in an area of just over 0.5 square meters. These detectors are used to identify short-lived, unstable particles like the bottom quark.
- The electromagnetic calorimeter (ECAL) is used to detect photons and electrons. It is made of lead tungstate crystals, which are 98 percent metal (by mass) but completely transparent. The 80,000 crystals in the ECAL have a total mass equivalent to that of 24 adult African elephants and are supported by 0.4-millimeter-thick structures made from carbon fiber (in the endcaps) and glass fiber (in the barrel) to a precision of a fraction of a millimeter.
- The hadronic calorimeter (HCAL) will be used to detect the energy from jets of particles. The brass used for the endcap of the HCAL comes from recycled artillery shells from Russian warships.

instruments such as cosmic ray detectors or radio telescopes look at the cosmos in different ways. Astronomy would be greatly impoverished if it had just one or two types of instruments. That is, different instruments can work in different ways to make discoveries that advance science.

Three types of instruments also can be identified in particle physics. First there are the proton accelerators, such as the Tevatron and the LHC, which offer the fastest route to the highest energy. They might be compared to very large ground-

FIGURE 3-1-2 In the underground tunnel of the LHC, the proton beams are steered in a circle by magnets. The LHC will provide particle collisions for the ATLAS and CMS experiments. Courtesy of CERN.

- The solenoid magnet, which allows the charge and momentum of particles to be measured, will be the largest solenoid ever built. The maximum magnetic field supplied by the solenoid is 4 tesla—approximately 100,000 times as strong as the magnetic field of Earth. The amount of iron used as the magnet return yoke is roughly equivalent to that used to build the Eiffel Tower in Paris. The energy stored in the CMS magnet when running at 4 tesla could be used to melt 18 tons of solid gold.
- During one second of CMS running, a data volume equivalent to the data in 10,000 Encyclopedia Britannicas will be recorded. The data rate to be handled by the CMS detector (approximately 500 gigabits per second) is equivalent to the amount of data currently exchanged by the world's telecommunication networks. (The data rate for ATLAS is similar.)

based telescopes. Second are the electron accelerators. At any point in history, the energy that was reachable with electron accelerators—such as those currently operating in California and Japan—has typically been lower than what could be reached with a proton accelerator, but electron collisions offer a much clearer picture of particle properties and interactions. Electron-positron colliders might be compared to HST. Finally, as in astronomy, there are a host of different instruments—nuclear reactors, underground laboratories, tabletop measurements, space-based observations, and more—each of which elicits entirely different kinds of information.

Science is full of uncertainty, and new discoveries from the LHC or elsewhere might change the picture. But as of today, the substantial majority of particle physicists in the United States, Europe, and Japan do not advocate that the next step in particle physics should be a larger facility of the same type as the Tevatron and the LHC. Rather, the dominant view—increasingly so in recent years—has been that the next step should be to push the frontier of clarity and sensitivity with a TeV-class electron-positron collider, the ILC. The initial phase of the ILC is envisioned to have a total energy of 500 GeV, with the possibility of a subsequent increase in the energy to 1 TeV.[1]

The ILC can make many important discoveries that are beyond the reach of the LHC, even though LHC energies will allow the production of particle states up to around 5 TeV. It can provide detailed information about phenomena that the LHC can only glimpse. These may include phenomena predicted in the Standard Model but not yet observed, such as the Higgs particle. They may include phenomena that are already observed but difficult to study fully at proton colliders, such as the top quark. Or they may include entirely new phenomena that emerge at the LHC, including supersymmetry, large extra dimensions, new particle forces, and more. The LHC can see farther (higher in energy) into the Terascale but with relatively blurry vision, while the ILC can see more clearly but not directly into the higher regions of the Terascale (see Figure 3-1).

The advantage of the ILC is that it collides electrons, which are simpler and easier to understand than the protons used at the Tevatron and the LHC. Protons can be accelerated more cheaply and easily, but electrons typically give more detailed information. In that respect, building the ILC will be like launching a telescope above Earth's atmosphere.

Historically, the energy reach of hadron colliders has been greater than that of electron colliders, while the ability to extract the details of collisions has been better with electron colliders than with hadron colliders. (For more discussion on this topic, see Box 3-2.) Most previous electron colliders accelerated the beams in circular orbits, allowing the beams to be reused again and again. Energy is lost in

---

[1]For a full description of the internationally agreed-upon general parameters for the ILC, please see International Linear Collider Steering Committee, Parameters Subcommittee, *Parameters for the Linear Collider*, September 2003; the report is available online at <http://www.fnal.gov/directorate/icfa/LC_parameters.pdf>. For the baseline configuration design of the ILC, please see <http://www.linearcollider.org/wiki/doku.php?id=bcd:bcd_home>.

FIGURE 3-1 As depicted in this artist's montage, while both the LHC (left) and ILC (right) will collide particles at Terascale energies, the character of the interactions will be quite different. For the LHC, protons (containing various elementary quarks) will collide; at the ILC, pointlike electrons (and positrons) will collide. Courtesy of CERN and DESY Hamburg.

each orbit of the electrons, however, and the energy loss increases dramatically as the energy of the beam is increased. For this reason, it is impractical to reach Terascale energies with a circular electron collider. To reach such energies in electron collisions requires the challenging new technology of a linear collider. An early accelerator of this type, the SLC, operated at the SLAC laboratory in California in the early 1990s and proved to be an important milestone in establishing the feasibility of a linear accelerator; the project also led to some of the most precise tests yet of the Standard Model (see Figure 3-2). Building on this experience and using novel technology, physicists today are proposing to build a large-scale version of an electron-positron linear collider—possibly 30 km long—that can explore the Terascale.

The LHC, with the high energy of its collisions, and the ILC, with the extremely precise measurements possible at an electron-positron collider, can combine to provide the necessary tools to explore the Terascale. Taken together, discoveries at the LHC and ILC could uncover the much anticipated mysteries of this new domain of nature.

BOX 3-2
Collisions of Different Types of Particles: Electrons vs. Protons

For a physicist, the electron is about as simple as a particle can be. It is called a "point particle," and it obeys the simplest laws that are allowed by the principles of relativity and quantum mechanics. Electrons have been smashed together at huge energies in accelerators and probed in ultraprecise tabletop experiments to measure their magnetic and electric properties. The results fit with the current understanding of the electron as a relativistic and quantum mechanical point particle.

The proton, by comparison, is not simple (see Figure 3-2-1). It is composed of simpler objects called quarks and gluons. The equations governing quarks and gluons have been known for 30 years, but they are so complex that even with modern supercomputers, physicists are still struggling to understand how quarks and gluons behave.

Electrons and protons, and their antimatter counterparts (the positron and antiproton), are the most easily accelerated particles. But they have contrasting virtues for experiments:

- Protons can be accelerated more easily than electrons to higher energies. Because proton accelerators can reach higher energies, they have been able to directly produce and discover heavy particles, including the W and Z particles and the top quark.
- The great advantage of electrons is that they are point particles. Collisions involving electrons are much easier to understand and interpret.

As a result, many discoveries have been made first with protons, and often the most precise measurements are made with electrons. For example, the direct evidence for quarks was demonstrated in electron-proton scattering experiments in the 1960s at SLAC. Proton-proton scattering had reached higher energies, but the results were too complicated to reveal the existence of quarks. More recently, many of the high-precision tests of the Standard Model have come from collisions involving electrons.

Physics at the Terascale

Discovering the Higgs Particle

According to the Standard Model, the difference between the weak interactions and electromagnetism is related to the origin of the masses of most elementary particles through the unusual behavior of a new particle called the Higgs particle. Whether this hypothesis is correct is not known experimentally. All that is

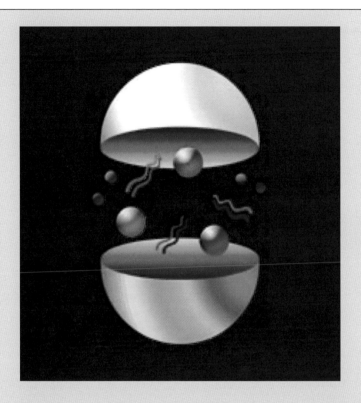

FIGURE 3-2-1 A proton consists mainly of three quarks, but it also contains gluons and other quarks and antiquarks, which makes it a very complex object. This artist's conception illustrates the nonelementary nature of the proton. Here the artist imagined cutting open a proton to see the material inside, including quarks (the three large balls), gluons (wiggly lines), and extra quark-antiquark pairs (the small balls that come in pairs).

known for sure, based on extrapolating from what has already been observed, is that at Terascale energies, either a Higgs particle will emerge or the Standard Model will become inconsistent and a new mechanism will be needed.

If the Standard Model is correct, the LHC will discover the Higgs particle. But its ability to test the Standard Model theory of the Higgs particle will be limited. Is the Higgs particle really responsible for particle masses? Have Higgs particle interactions hidden the weak interactions from our everyday experience, as the Standard Model claims? Is there just one Higgs particle, or several? Answering these

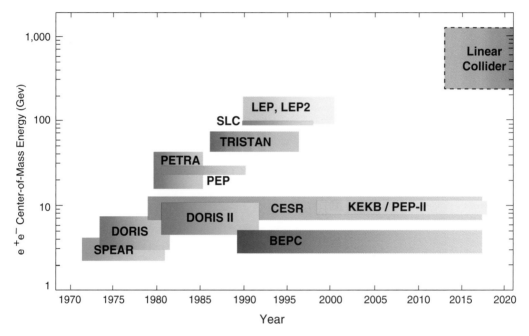

FIGURE 3-2 A 30-year history of electron colliders around the world indicates the increasing energy of collision. The colored bars represent the chief operating periods of the named accelerators. The red region in the upper right corner surrounded by a dashed line represents a proposed scenario for the ILC. Figure content courtesy of M. Tigner, Cornell University, and R.N. Cahn, Lawrence Berkeley National Laboratory.

questions requires measuring the interactions of Higgs particles in a more precise way than can be done at the LHC. The high energy of the LHC will enable it to produce and detect Higgs particles if the Standard Model is correct, but the complexity of proton interactions limits the information about these particles obtainable from the LHC.

The ILC will be able to zoom in on the Higgs particle and measure its properties and to measure multiple Higgs particle interactions with high precision. The ILC will be sensitive to subtle modifications of the behavior of the Higgs particle resulting from unknown physics at much higher energies, perhaps even from exotic new physics such as extra dimensions of space and time (see Figure 3-3).

Of course, it is possible that the Standard Model theory of weak interaction symmetry breaking and particle masses is incorrect, or not entirely correct. Perhaps instead of a Higgs particle there is a more exotic mechanism behind these phenomena—possibly something that physicists have not even thought of yet. Or perhaps something exists that is somewhat like a Higgs particle but the Standard

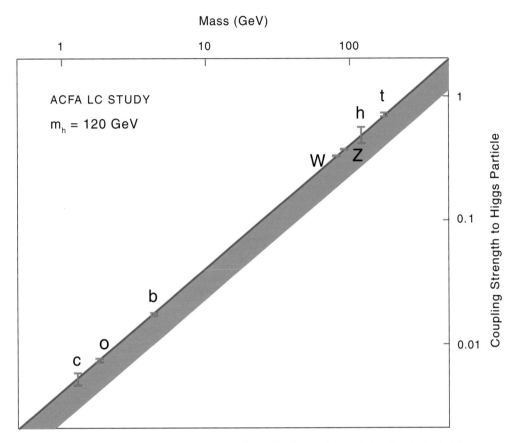

FIGURE 3-3 The interactions of the Higgs particle with the particles of the Standard Model will generally be sensitive to the presence of extra dimensions of space and time. The orange error bars show the precision possible at an ILC for the measurements of the couplings of the Higgs particle to other particles, while the green band shows the range of predictions in theories with extra spatial dimensions. The Standard Model prediction is the upper edge of the green band. If extra special dimensions exist, the measurements of the Higgs couplings obtained at the ILC could provide evidence for them. Courtesy of American Linear Collider Physics Group.

Model does not describe it correctly. In any case, data from the LHC may be confusing, difficult to interpret, or subject to misinterpretation. The greater clarity and precision of the ILC will likely be even more important if the Standard Model theory of these phenomena is incomplete or incorrect.

Even such a basic property of the Higgs particle as its spin cannot be easily measured at the LHC. The Standard Model requires that the Higgs particle has no spin (in contrast to, say, the electron and proton, which spin like tiny magnets). If a Higgs particle is discovered, the spin can be measured straightforwardly by determining the rate at which it is produced at different energies at the ILC.

The precision measurements at a linear collider together with the results from LHC are crucial to establish the Higgs mechanism responsible for the origin of mass and for revealing the character of the Higgs boson. If the electroweak symmetry is broken in a more complicated way than foreseen in the Standard Model, the ILC and LHC together can help define alternative models of Terascale physics.

Supersymmetry: The Search for New Quantum Dimensions

Past measurements of particle interactions have given hints that a new phenomenon known as supersymmetry might emerge at the Terascale. Supersymmetry, if it is correct, updates Einstein's theory of special relativity by including quantum variables in the description of space and time. Ordinary dimensions are measured by numbers—it is 3 o'clock, we are 200 meters above sea level at 40 degrees north latitude, and so on. If nature is supersymmetric, space and time will have new quantum dimensions as well as the familiar dimensions that we see in everyday life.

Previous hints for the existence of supersymmetry come from two types of measurements. First, based on the rates that are measured for the different particle interactions, it appears that the particle forces all have equal strength at very high energies if nature is supersymmetric; otherwise, they differ by small amounts. Second, supersymmetry gives a satisfactory explanation of why observed particle masses are so tiny compared to the energy of particle unification, which is expected to be around 10^{16} GeV. The inability to explain this disparity is considered a serious drawback of the Standard Model.

What might be observed in the laboratory if supersymmetry is correct? Vibrations of ordinary particles in the new supersymmetric dimensions will give rise to new particles with distinctive properties. In a supersymmetric world, there are supersymmetric "shadows" of the known particles—a little like the shadow world of antimatter that was discovered in the 20th century. These new particles are called superpartners and may well provide the explanation for dark matter (see Figure 3-4).

If supersymmetry becomes apparent at the Terascale, the LHC will blaze the first trail. It will discover some and possibly many of the superpartner particles and make numerous important measurements. But many of the most important measurements will be out of reach of the LHC. Physicists will need the ILC to make the crucial measurements to verify that the new particles are indeed supersymmetric counterparts of the observed particles, to understand their main properties, and possibly to gain a new understanding of the unification of forces (see Figure 3-5).

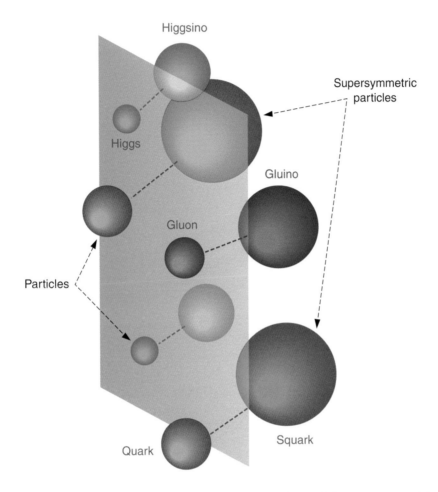

FIGURE 3-4 A key feature of supersymmetry is the existence of so-called superpartner particles for each type of ordinary particle.

Extra Dimensions of Space and Time

Supersymmetry is by no means the most exotic possibility for physics at the Terascale. Space may have extra dimensions beyond the three that we experience in everyday life. These are new dimensions that would be unlike the quantum dimensions of supersymmetry; they would be more akin to ordinary dimensions, like the ones seen in everyday life except smaller. These extra dimensions sound like science fiction, but they are the basis for fascinating theories of physics at the Terascale (see Figure 3-6).

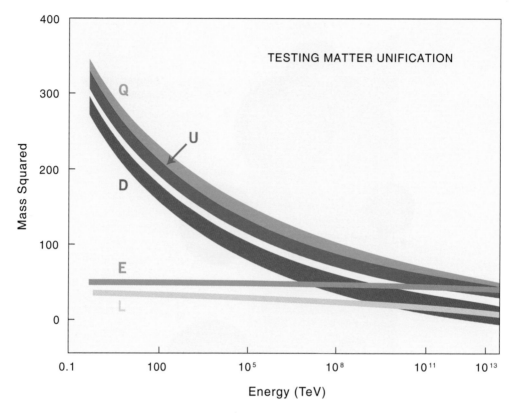

FIGURE 3-5 Particle masses depend on the energy at which they are measured. If superpartner masses are measured (the left-hand edge of the plot) at the LHC and ILC, then their masses at very high energies can be calculated and used to test the theory of unification. The Q, U, and D curves are the masses of the superpartners of the quarks measured at the LHC, while the E and L curves are the masses of the superpartners of the electron and neutrino measured at the ILC. The bands represent the potential experimental accuracy. This test of unification requires both the LHC and the ILC. Courtesy of the American Linear Collider Physics Group.

These extra dimensions, if they exist, must be small, simply because they have not yet been detected. Discovering such extra dimensions requires the high energy of particle accelerators. If the extra dimensions are large enough, the LHC will obtain the first indication that they exist by observing and studying collisions in which energy seems to disappear. If such events are seen, there will be many possible explanations, including the disappearance of the missing energy into new dimensions of space.

Learning if this is the right explanation will take a great deal of work, and the

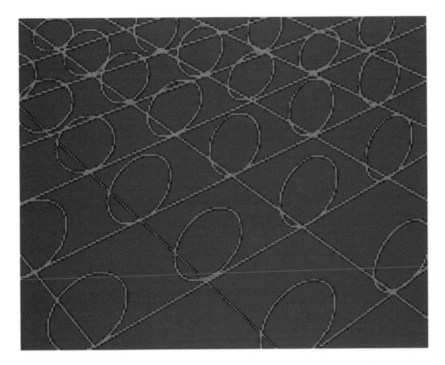

FIGURE 3-6 Artist's conception of small extra dimensions proposed in superstring theories. The circles represent an additional spatial dimension that is curled up within every point of familiar three-dimensional space; shown here is a two-dimensional space (the plane of intersecting lines) with a third dimension that is small because it is curled up (shown as a loop).

LHC can make some of the important measurements. The ILC, however, will be able to go much farther, measuring many new properties of exotic events and gaining far more information on the number and shape of possible extra dimensions of space (see Figure 3-7). Its ability to do this depends on the fact that in an electron collider one can control the energy of the incoming electron beams. A proton collider does not have the same degree of control because a proton is made of many quarks and gluons. In a proton collider, the energies of the quarks and gluons responsible for a specific high-energy collision can vary over a large range.

Dark Matter

One of the great surprises in astronomy is that matter of the sort familiar to us—atoms and molecules, electrons, protons, and neutrons—makes up only about

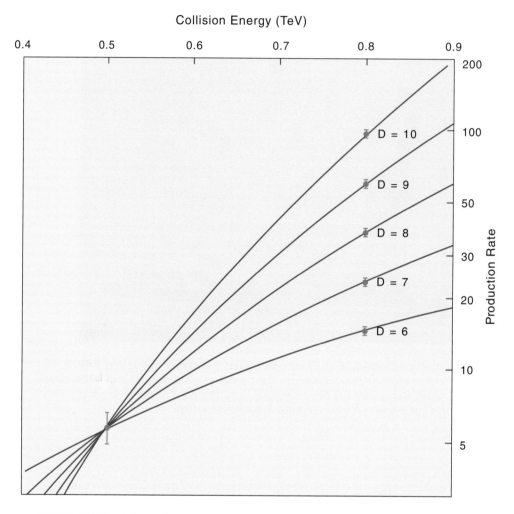

FIGURE 3-7 The ILC can observe particles that seem to disappear into extra dimensions, because such events appear not to conserve energy as particles disappear into the extra dimensions. The production rate for this type of process depends on the incoming electron beam energy and the number and size of the extra dimensions. In this example, the size of the extra dimensions has been chosen so that all the curves overlap at 500 GeV. The different lines on the graph are indexed by the number, n, of extra dimensions, D. It is possible that the production rates will overlap at one energy, as shown in the diagram at 0.5 TeV; however, by running the ILC at more than one energy, the number of extra dimensions can be determined. The capability of the ILC to change the collision energy of the electrons is thus crucial to this type of measurement. Courtesy of the American Linear Collider Physics Group.

4 percent of all the matter in the universe. The rest is dark matter, inferred from its gravitational effects but not observed, or dark energy.

What is dark matter? Calculations suggest that it consists of Terascale particles, though these guesses require physics beyond the Standard Model. No suitable particle is predicted in the Standard Model, and none has been observed so far. Thus, dark matter almost certainly consists of particles that do not exist in the Standard Model and whose nature and origin are a mystery.

The current understanding of particle physics can be extrapolated to very early times, shortly after the big bang, when the universe was dense and hot and all particles were in thermal equilibrium. Then, as the universe expanded and cooled, most of the dark matter particles decayed and disappeared. How much dark matter one ends up with now depends on the mass and interaction rates of dark matter particles. Terascale particles (such as new particles associated with supersymmetry) turn out to have just about the right properties.

Physicists are now looking for dark matter particles in experiments placed deep underground to shield them from ordinary cosmic rays. If the dark matter is really made of Terascale particles, there is a good chance of detection within the next decade. However, dark matter observatories, if they find something, will not quite be able to determine what they have found. They will reveal something about the mass, the abundance in the universe, and the interaction rates of the dark matter particles, but they will not produce enough information to disentangle these properties and piece together the whole story.

To understand Terascale dark matter and its role in particle physics, there is no good substitute for actually producing it and studying it in the lab (see Figure 3-8). The LHC has an excellent chance of making the first observation. By studying how energy and momentum seem to disappear when dark matter particles are created, physicists working at the LHC should be able to make an initial discovery. The ILC then would serve as the ideal dark matter microscope. By making the detailed measurements that show how much a new particle contributes to dark matter, the ILC could precisely determine how many of the Terascale particles should be left over from the big bang. These measurements have profound implications for both particle physics and cosmology.

The Standard Model and Beyond

The ILC will probe the Standard Model with unprecedented precision, well beyond what was achieved in the last decade by electron colliders of lower energy at the SLAC and CERN laboratories in California and Europe, respectively. Literally dozens of high-precision measurements will be made, involving the masses, lifetimes, and reaction rates of W and Z particles, top quarks, possibly Higgs

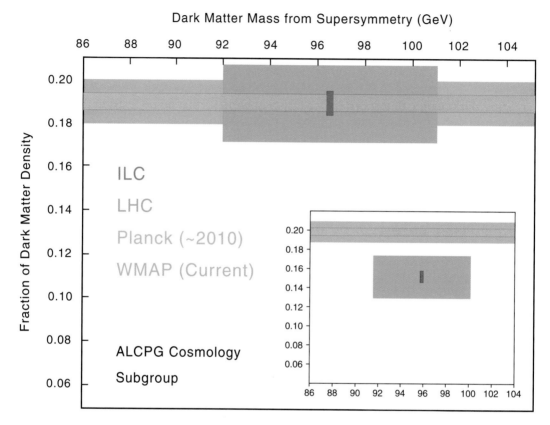

FIGURE 3-8 The WMAP and Planck satellites may determine the total amount of dark matter in the universe, but they will not measure the dark matter's mass. The LHC and ILC colliders could determine the mass of an individual dark matter particle. For example, agreement between satellite and collider measurements might imply that supersymmetric particles known as neutralinos are the dark matter. As shown in the full diagram, the ILC would offer substantially improved measurement precision in this comparison. Potential disagreement, as shown in the inset, would provide evidence for additional dark matter components. Courtesy of the American Linear Collider Physics Group Cosmology Subgroup.

particles, and others. If the Standard Model survives these tests, physicists will gain a new level of confidence in its validity and scope.

There are many ways in which the Standard Model could fail. The ILC can infer the validity or breakdown of the Standard Model even at energies beyond the Terascale. This is because quantum mechanics allows new particles to appear briefly, influencing a reaction among lighter particles, even if there does not seem to be enough energy for the heavier particles to play a role. This facet of the quantum world is a fundamental characteristic of nature, being responsible, for example, for the radioactive decay of certain atomic nuclei. Taking advantage of this, physicists using the LHC, and especially the ILC, will be able to explore the validity of the Standard Model at energies even higher than can be reached with today's technologies.

Toward the Terascale

Soon the LHC will begin the exploration of the Terascale, and a proposed linear collider would extend this exploration into unknown realms and add new insights to those discoveries. Together, these two accelerators would enable physicists to probe the critical questions of particle physics in many different ways. The exploration of the Terascale with the LHC and the ILC is the top scientific challenge of particle physicists today. As such, direct investigation of this energy frontier continues to offer the broadest approach to the questions posed in the previous chapter.

The LHC will be ready soon, whereas the international effort to design the ILC is still under way, relatively speaking. Invoking the analogy of exploring a new landscape, the LHC can provide a bird's-eye view of the most interesting features. The ILC will be able to focus on specific landmarks with exquisite precision as well as with different observational capabilities.[2] Table 3-1 provides some specific examples of the combined discovery potential of the LHC and the ILC.

The ILC may be able to discover important but rare or hard-to-detect processes that the LHC will miss (see Box 3-3). For example, the ILC will be able to measure the relevant quantum numbers and lifetimes of the particles that it detects. To take a recent example, more than a decade after the discovery of the top quark at the Tevatron, very little is known about this particle. Its lifetime, its spin, and even its charge have not been experimentally determined. Physicists are sure that it is the top quark because its mass and a few of its key properties have been measured, and they agree with the Standard Model expectations. Moreover, the Standard Model is extremely successful, and its applicability to quarks is well established. Thus, measuring the top quark mass and some of its properties was enough to claim a discovery.

However, discoveries at the LHC are likely to be a different matter, especially

---

[2]The committee acknowledges its indebtedness to HEPAP for providing this useful analogy. (See HEPAP, *Discovering the Quantum Universe*.)

TABLE 3-1 Potential Synergies Between the ILC and LHC in Explorations of the Terascale

If LHC Discovers:	What ILC Could Do:
A Higgs particle	Discover why the Higgs exists and who its cousins are. Discover effects of extra dimensions or a new source of matter-antimatter asymmetry.
Superpartner particles	Detect the symmetry or supersymmetry. Reveal the supersymmetric nature of dark matter. Discover force unification and matter unification at ultra-high energies.
Evidence for extra dimensions	Discover the number and shape of the extra dimensions. Discover which particles are travelers in extra dimensions and determine their locations within them.
Missing energy from a weakly interacting heavy particle	Discover its identity as dark matter. Determine what fraction of the total dark matter it accounts for.
Heavy charged particles that appear to be stable	Discover that these eventually decay into very weakly interacting massive particles; identify these "super WIMPS" as dark matter.
A Z-prime particle, representing a previously unknown force of nature	Discover the origin of the Z-prime. Connect this new force to the unification of quarks with neutrinos, or quarks with the Higgs, or with extra dimensions.
Superpartner particles matching the predictions of supergravity	Discover telltale effects from the vibrations of superstrings.

if there is a major breakdown of the Standard Model at the Terascale. If a particle is discovered near 135 GeV, for example, one might suspect that it is the long-sought-after Higgs particle, but one could not be certain just from the initial observation of the particle. Because the Standard Model theory of Higgs particles has not been tested, it will be necessary to measure all of the properties of the purported Higgs particle. The LHC will begin this job, and the ILC will continue it. The ILC will have the capability to measure key quantum numbers, coupling constants, and lifetimes in a way independent of specific models.

There have been two distinct and complementary strategies for gaining new understandings of energy, matter, space, and time at particle accelerators:

- Exploration of new energy regimes to directly discover new phenomena, such as by using accelerators operating at the energy scale of the new particle.
- High-precision measurements to observe differences in expected patterns of behavior to infer new physics—that is, searching for the quantum echoes of higher energy phenomena.

These two strategies have worked well together, revealing much more than either by itself. Enough is now known to predict with confidence that a linear collider will be needed to fully answer the key questions about ultimate unification, the origin of mass, the nature of dark matter, and the structure of space and time that the LHC will begin to address.

HIGH-INTENSITY BEAMS

Some questions in particle physics are answered best not by the highest energy beams but by very intense beams, such as intense sources of bottom quarks or beams of neutrinos (see Boxes 3-4 and 3-5). These beams are valuable because they can reveal processes that occur very rarely. Also, very intense beams, like high-energy beams, offer a window onto energies that are beyond the reach of accelerators through the small but perceptible effects of very massive particles on low-energy processes. In addition, intense beams are needed to study neutrinos, since a vanishingly small percentage of neutrinos leave a trace in a typical detector.

The B factories that produce bottom quarks in abundance are one example of high-intensity beams. By the end of this decade, the B factories at KEK in Japan and SLAC in California will have observed billions of B meson decays, in addition to the B meson decays observed at Cornell's CESR accelerator. These decays have provided a solid understanding of charge parity (CP) violation as it affects quarks. They also have allowed physicists to explore indirectly (and rule out) some of the phenomena hypothesized for the Terascale.

If the LHC sees phenomena that are inexplicable within the Standard Model, such as the particles associated with supersymmetry, studies of B meson decays could reveal some of their properties. A relative of the B meson, called the B_s meson, has been produced in sufficient quantity for detailed studies at hadron colliders. The study of the B_s meson has begun at Fermilab and will be expanded in the next decade with the LHCb experiment now under construction at CERN. A super-B factory might expand on the sample of B meson decays by as much as an additional factor of 10, allowing the measurement of even rarer events. Ideas for such a facility are being studied in both Japan and Italy.

Intense beams of neutrinos allow physicists to study neutrino oscillations, in which neutrinos of one variety morph into another variety as they travel. The most incisive information about neutrino oscillations today comes from experiments using neutrinos from a variety of sources: nuclear reactors, the sun, cosmic ray interactions in the upper atmosphere, and accelerators. Experimental observations of oscillations from atmospheric neutrinos have been verified using accelerator-produced neutrinos from the KEK accelerator in Japan (the K2K experiment) that mimic those from the atmosphere. Two accelerator experiments are now in

BOX 3-3
The Science of the ILC

Exploration of the Terascale will be at the center of particle physics in the coming decades. The journey will begin with the discoveries of the LHC and will continue with the ILC, a proposed new accelerator designed to make discoveries at the Terascale and beyond.

The ILC will consist of two linear accelerators, each about 15 kilometers long, aimed straight at each other (see Figures 3-3-1 and 3-3-2). One accelerator will contain electrons, the other positrons. The electrons and positrons will be assembled into bunches, each containing 10 billion particles. The particles will be accelerated to near the speed of light and then brought into collision. At the collision point, the beams will be focused down to approximately 3 nanometers wide. In the resulting collisions, electrons and positrons will annihilate into energy and produce new particles.

Electron-positron collisions are clean and precise, and in a linear collider the energy can be adjusted to focus on the physics of interest. For the ILC, the plan is to start at a center-of-mass energy of 500 GeV, with a later increase in energy to 1 TeV. The initial energy is sufficient to produce and study Higgs particles and possibly other new physics; the higher energy might well be needed to access additional new features of the Terascale. The timing and nature of the energy upgrade will depend on what is found at the LHC and on the ILC's initial operation.

The scientific case for the ILC has many components:

- If the LHC finds a new particle, the ILC will be necessary to measure its properties precisely and determine definitively whether it is the predicted Higgs particle.
- If the universe is supersymmetric, the LHC and ILC will both be needed to discover and understand the new world of superpartners predicted in these theories.
- The ILC could study the properties of the lightest superpartner with great precision to determine whether it makes up some or all of the dark matter.
- The LHC and the ILC will also address many questions about extra dimensions. Does the universe have more dimensions than those we observe? The LHC can find evidence for the existence of hidden dimensions; the ILC can map their nature, shapes, and sizes.

In whatever direction the LHC points, the ILC will push even farther the exploration of the mysteries of the Terascale.

progress at Fermilab. MINOS is a long-baseline experiment that will precisely measure the difference in neutrino masses corresponding to the atmospheric neutrinos studied at K2K.[3] The MiniBoone experiment will confirm or rule out a

---

[3]First results from the MINOS experiment have been announced that are consistent with the K2K and Super-K measurements. For more information, see the press release at <http://www.fnal.gov/presspass/press_releases/minos-3-30-06.html>.

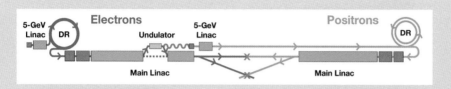

FIGURE 3-3-1 A schematic layout of the ILC. This diagram reflects the recommendations of the Baseline Configuration Document, a report published in December 2005 that outlines the general design of the machine. Courtesy of the ILC Global Design Effort.

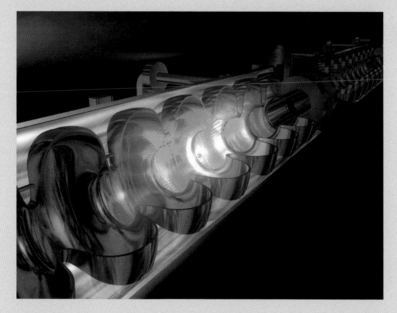

FIGURE 3-3-2 Artist's conception of the ILC accelerator structure in the underground tunnel; the cutaway view shows the interior of the superconducting cavities. Courtesy of DESY Hamburg.

result from a still controversial experiment, which used the meson facility at Los Alamos National Laboratory, suggesting that there may be more than three neutrinos. In Europe, the CERN-Neutrinos-to-Gran-Sasso experiment will be operational by late 2006 and will seek to directly observe the oscillations of the muon-type neutrino over a distance of 730 km.

Neutrino experiments have demonstrated that any one neutrino of definite mass can be thought of as a quantum mechanical mixture of the electron neutrino,

BOX 3-4
Flavor Physics: Precision Science of Particle Interactions

Progress in particle physics at the energy frontier has been complemented by precision investigation of flavor physics—studies of the patterns of weak decays. In the Standard Model, these decay patterns for quarks are predicted by a set of four parameters that define a set of couplings known as the Cabibbo Kobayashi Maskawa matrix. A single one of these four parameters accounts for all CP-violating effects, which produce the differences between the laws of physics for matter and antimatter. Physicists are interested in studying such effects because the existence of matter in the universe is thought to depend on this difference.

Once the Standard Model theory is extended to include neutrino masses, there is an additional matrix of parameters relating the light leptons—the electron, muon, tau, and their neutrinos. Little is known today about the details of these parameters, which also include additional possibilities for CP violation and thus provide another possible, and quite different, root cause for the matter-antimatter imbalance in the universe.

Because there are few parameters describing all weak decays of quarks and a similarly small set describing decays that change one family of lepton into another, the Standard Model can be subjected to precision testing in this sector. Any new particles, even with masses beyond the range of current accelerators, can contribute to these decays through unseen quantum intermediate states. In many cases, such contributions would be detectable because they destroy the decay patterns predicted by the Standard Model alone.

The effect of such new particles decreases rapidly if the mass of the new state is larger. Therefore, sensitivity to discovering these unseen but indirectly involved particles is greatest when particle physicists have very accurate knowledge of the Standard Model prediction for a specific experiment. In particular, where the new contribution makes possible a decay that was predicted to be extremely rare (or even absolutely forbidden) in the Standard Model, very sensitive searches can be made for the indirect effects of new heavy particles. These precision measurements provide a window on new physics that can in some cases be as sensitive as direct searches at high energy.

The observations from flavor physics are complementary to the capabilities of both the LHC and ILC and hence will continue to provide important information, even when these facilities begin to probe the Terascale directly. The LHC beauty experiment (LHCb) experiment will probe some of this physics, and a possible super-B factory experiment can add studies of modes that are very difficult to study in the LHC environment. Experiments to search for lepton flavor violations and further study of the neutrino "mixing" matrix provide a separate opportunity for study.

muon neutrino, and tau neutrino. In this way, neutrinos produced as a definite flavor type can oscillate or mix to become a different type as they travel. The basic mixing is reminiscent of the pattern already known for quarks, but the mixing effect is small for quarks and surprisingly large for neutrinos. Experiments have measured two of the three parameters that describe how neutrino mixing occurs. The third mixing parameter, known as θ_{13}, is not as large as the other two and so far has eluded experimenters.

The next generation of experiments, including possible reactor experiments,

BOX 3-5
Neutrinos: An Enigma Wrapped in a Mystery

Neutrinos are among the least understood of the fundamental particles. They are similar to the more familiar electron, with one crucial difference: Neutrinos do not carry electric charge. Because neutrinos are electrically neutral, they are not affected by the electromagnetic forces that act on electrons. Neutrinos are affected only by the weak force, which has a much shorter range than electromagnetism. They therefore are able to pass through great quantities of matter without being affected by it. It would take a wall of ordinary matter more than 100 light-years thick to stop a beam of neutrinos like those produced by the sun. Precisely because they are so elusive, neutrinos produced at the center of the sun traverse the entire mass of the sun without being absorbed, providing a way to see deep into the sun's interior.

John Updike's 1959 poem "Cosmic Gall" featured neutrinos' two most important and puzzling features—masslessness and elusiveness. Today, it is known that neutrinos are almost, but not quite, massless. However, even by subatomic standards, neutrinos have only minuscule masses and are therefore only barely affected by gravity.

Three types of neutrinos are known; there is strong evidence that no additional neutrinos exist, unless their properties are unexpectedly very different from the known types. Each type, or flavor, of neutrino is related to a charged particle (which gives the corresponding neutrino its name). Hence, the electron neutrino is associated with the electron, and two other neutrinos are associated with heavier versions of the electron called the muon and the tau (see Figure 3-5-1).

Experiments are needed to complete the picture. The pattern of partnerships is determined by the ordering of the masses, and it is not yet known whether the electron associates with v_1 or v_3 (an issue known as the hierarchy problem). The picture that is emerging is reminiscent of the pattern for quarks in the weak interaction, but the effect is much more dramatic for the leptons because neutrino mixing is a much larger effect.

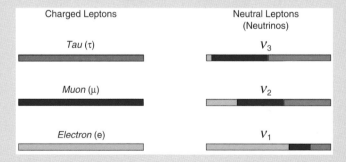

FIGURE 3-5-1 Schematic depiction of how the neutrinos fit into the new version of the Standard Model along with their charged lepton partners, the electron (e), muon (μ), and tau (τ). The colored segments represent the relative proportions in which each particle incorporates the property that characterizes it as electron, muon, and tau in the weak interaction. The electron, muon, and tau each have single colors and are states with definite mass. The observed partner neutrino particles are v_1, v_2, and v_3; they are multicolored, indicating that each is a mixture of the neutrino flavor states v_e, v_μ, and v_τ.

Continued

BOX 3-5 Continued

Neutrinos have been shown to oscillate, which demonstrates, in effect, that they have mass. Understanding neutrino oscillations requires a trip into the world of quantum mechanics (see Figure 3-5-2).

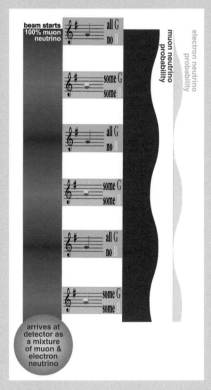

FIGURE 3-5-2 The image uses a musical analogy to represent the behavior of a simplified model. Imagine two neutrinos that can oscillate into one another, and imagine representing each neutrino as a musical pitch. Further assume that only one pitch at a time can be detected. Let the muon neutrino be represented by a G-note and the electron neutrino by, say, a B-note. In the absence of neutrino oscillations, one could assume that a G originated as a G and would remain forever a G, and the same would be true of a B. However, with the possibility of neutrino oscillations, a muon neutrino G can "de-tune" into a B as time passes, and vice versa. Since only one pitch at a time can be detected, the neutrino will sometimes sound like a G and sometimes like a B; the rate of detuning is related to the neutrino mixing parameters. The probability of observing the muon neutrino as an electron neutrino varies as a function of time (or distance if the neutrino is traveling), as shown by the sinusoidal curves alongside the scales. The detailed properties of neutrino oscillations are important to understanding how Standard Model particles interact and the properties of galaxies and the universe. Courtesy of Paul Nienaber and Andrew Finn, BooNE Collaboration.

the NOνA experiment at Fermilab, and the Tokai-to-Kamioka (T2K) experiment at J-PARC, hope to measure the neutrino-mixing parameter θ_{13}. The proposed NOνA experiment at Fermilab would not only be sensitive to θ_{13} but may also be able to use the interaction of neutrinos with Earth to learn whether neutrinos masses are ordered in a way reminiscent of quarks and the charged leptons. The ordering of the neutrino masses could be a critical clue for understanding what the structure of the constituents of the Standard Model reveals about the underlying physics. Taken together, the proposed NOνA and T2K experiments would reveal somewhat more information than either one alone.[4] The amount of additional information gained by carrying out both experimental programs depends critically on the value of θ_{13}. If θ_{13} is too small, the planned experiments will not be sensitive to the neutrino mixing and will have more limited scientific value. The ultimate goal of this line of research is to understand the possible pattern of CP violation in the neutrino sector, which might have contributed to the dominance of matter over antimatter in the early universe as revealed by astrophysical observations.

If θ_{13} is big enough, experiments might be able to detect CP violation in neutrinos. The most sensitive searches for both θ_{13} and CP violation will require massive detectors and extremely intense beams of neutrinos. The United States is investigating possible designs for a facility that would produce neutrinos from an intense beam of protons known as a "proton driver." Japan currently has such a facility under construction at J-PARC. In the longer run, ultrapure beams of electron neutrinos produced either from the radioactive decay of beams of unstable atomic nuclei ("beta beams") or from a neutrino factory might be required to pin down the issue of CP violation in neutrinos (though a realistic design for a neutrino factory is at least a decade away). A future generation of neutrino experiments may require underground detectors much more massive then the ones that already exist. Ironically, what became the first underground neutrino detectors were originally motivated by the hope of discovering that the proton is not stable. Proton decay is expected in many unified theories, and in many models the predicted proton lifetime is very near the current experimental sensitivity. Observing proton decay would be a major step forward in particle physics. In any case, future massive underground neutrino detectors can also serve as much more sensitive experiments to discover proton decay.

---

[4]See, for example, the recent report of the U.S. Neutrino Science Assessment Group to the DOE/NSF High Energy Physics Advisory Panel and the DOE/NSF Nuclear Science Advisory Committee in February 2006; available online at <http://www.science.doe.gov/hep/HEPAP/Mar2006/NuSAG_to_HEPAP_mar06.pdf> (last accessed March 10, 2006).

Other experiments would use intense beams of muons or K mesons. These experiments study very rare processes in the decays of K mesons that would pin down the underlying parameters that govern the Standard Model. Experiments like these are being proposed for J-PARC, CERN, and the Paul Scherrer Institute in Switzerland. A source of ultracold neutrons is planned at several places in the world to measure the electric dipole moment of the neutron with a sensitivity two orders of magnitude greater than existing limits. A finite value would signal time-symmetry violation beyond that expected in the Standard Model and could help to explain the dominance of matter over antimatter in the universe.

NATURE'S PARTICLE SOURCES

Nature also produces particles. Gamma ray photons or neutrinos from outer space can have very high energies. The background buzz of particles traversing the galaxy or universe can serve as a sort of astrophysical laboratory, reflecting the ongoing evolution of the universe (see Box 3-6). A slab of solid material or a volume of liquid or gas of terrestrial origin can serve simultaneously as a source of particles (via decay of the constituents, such as by radioactive decay) and as a detector of particles (by providing a detecting medium for interactions of external particles with the constituents of the material).

For example, the radioactive decays of nuclei provide some information about the mass of the electron neutrino. When a neutron converts to a proton in nuclear beta decay, an electron is released whose characteristic energy spectrum depends, ever so slightly, on the mass of the electron neutrino. If the neutrino mass is large enough, this distortion will be visible. Sensitive experiments using the beta decay of tritium have been carried out for many years. Using the most ambitious experiment so far conceived, an international collaboration is mounting an experiment in Germany called KATRIN, which is designed to be sensitive to the distortion from a neutrino with a mass less than 1 eV.

Double beta decay, which was first observed in 1986, is a radioactive process in which two neutrons in the same nucleus simultaneously convert to protons, emitting two electrons and two neutrinos. However, many physicists suspect that a rarer and not yet observed type of double beta decay can also occur. If the neutrino is its own antiparticle, it is possible to have a double beta decay process in which no neutrinos are emitted at all. If such neutrinoless double beta decay is observed, the understanding of neutrinos will change substantially. This discovery would show that the origin of neutrino masses is very different from that of the masses of other known particles. It might provide a way to measure for the first time the overall mass scale of the neutrinos, and it might give a glimpse into physics at energies far beyond the Terascale, possibly involving particle unification.

BOX 3-6
Experimental Frontiers of Particle Astrophysics

Experiments in particle astrophysics use a remarkable range of techniques to address fundamental questions about the composition and evolution of the universe. In addition to searching for new physics, the experimental approaches listed here will help narrow uncertainties about fundamental parameters of cosmology:

• *Probing the nature of dark matter.*
  —Direct detection of dark matter particles in the Milky Way passing through Earth,
  —Direct production of dark matter particles in accelerators,
  —Detection of gamma rays from dark matter particle annihilations in the cores of galaxies, in dark matter clumps, and in the sun and Earth,
  —Improved observations of dwarf galaxies and small-scale structure to study clustering of dark matter and to test alternative models for dark matter, and
  —Measurement of the CMB temperature anisotropy and large-scale structure of the CMB to search for new particles that may contribute to a portion of the dark matter.

• *Testing cosmological models and probing new physics.*
  —Measuring CMB polarization to test inflationary theories (versus alternative cosmologies) and to find evidence for new physics at energies much greater than the Terascale (10 billion times greater),
  —Direct detection of gravity waves to probe new physics at scales between the Terascale and the inflationary scale,
  —Improved tests of general relativity to search for effects of extra dimensions or string theory,
  —Long-wavelength radio studies of 21-cm radiation from the early universe to probe cosmic evolution,
  —Measuring the time variation of physical constants using spectroscopy of distant objects to search for effects of extra dimensions and string theory,
  —Observing neutrinos and cosmic rays to understand the high-energy astrophysical sources that generated them, and
  —Using observations of neutrinos generated in the sun to better understand the solar core and the properties of neutrinos.

• *Probing the nature of dark energy.*
  —More accurate measurements of distances to and redshifts of supernovae to measure the dark energy equation of state,
  —Optical maps of gravitational lensing to determine the effect of dark energy on the growth of structure in the universe,
  —Measuring large-scale structure and baryon acoustic oscillations with redshift surveys to measure the dark energy equation of state, and
  —Observation of CMB temperature and polarization using satellites and ground-based experiments to precisely measure the amount of dark energy and to search for spatial nonuniformities in its distribution.

TABLE 3-2 Potential Neutrinoless Double-Beta Decay Experiments

Experiment	Isotope	Sensitivity (meV)		Comments
		Near Term	Upgrade	
CUORE	^{130}Te	189	63	Total mass = 750 kg; upgrade isotope purity
EXO	^{136}Xe	330	59	Upgrade total mass from 200 kg to 1,000 kg
Majorana M180	^{76}Ge	130		
MG1000			51	Hypothetical total mass 1,000 kg
MOON	^{100}Mo	403	141	Upgrade total mass from 200 kg to 1,000 kg
	^{82}Se	141	34	and run longer
Super-NEMO	^{82}Se	153		Total mass = 100 kg

NOTE: The exact scenario will depend on the real physics of our universe, but these examples give a taste of the potential of combining these two tools for exploration. SOURCE HEPAP, *Discovering the Quantum Universe*, 2006. For six selected neutrinoless double beta decay experiments, the signal sensitivities for neutrino mass in units of millielectronvolts (meV) are shown for a first-stage experiment as well as an upgraded capability. Different approaches use different radioactive isotopes (denoted by chemical symbol and total number of nucleons) to generate the beta decays. SOURCE: Neutrino Scientific Assessment Group; *Recommendations to the Department of Energy and the National Science Foundation on a United States Program in Neutrinoless Double Beta Decay*, September 1, 2005.

These experiments are notoriously difficult because radioactive decays from trace contamination in the material or the surroundings can produce a false signal. To overcome this liability, these experiments use ultrapure materials cooled to suppress background events and are located deep underground, which greatly reduces rates for cosmic ray events. A number of experiments are under way or planned to look for these phenomena (see Table 3-2).

The existence of dark matter was first inferred in the 1930s by measuring the motions of galaxies in large clusters.[5] However, the identity of the dark matter has remained a mystery. If dark matter is composed of weakly interacting elementary particles, as many astrophysicists and particle physicists believe, then, as Earth passes through a cloud of dark matter in its path around the sun, some of these particles can easily pass through the atmosphere and thousands of feet of rock to reach a detector deep underground. As they travel through the detector, it is expected that some will occasionally scatter off an atomic nucleus, causing the nucleus to recoil with the energy of a few tens of thousands of electron volts.

[5]This section identifies only a few of the ongoing and planned experiments in particle astrophysics. A more complete list (as of 2003) can be found in NRC, *Connecting Quarks with the Cosmos: Eleven Science Questions for the New Century*, Washington, D.C.: The National Academies Press, 2003; and DOE, Scientific Assessment Group for Experiments in Non-Accelerator Physics, available at <http://www.science.doe.gov/hep/SAGENAPFINAL.pdf>.

Detectors such as the Cryogenic Dark Matter Search (CDMS) in an abandoned iron mine in Soudan, Minnesota, use germanium and silicon detectors to detect such processes. Other examples include the Zoned Electroluminescence and Primary Light in Noble Gases (ZEPLIN) experiments in Britain's Boulby mine, which detects the light produced when a nucleus recoils in liquid xenon; the WIMP Argon Program (WARP), which uses liquid argon; and the Directional Recoil Identification from Tracks (DRIFT-II) experiment, also in the Boulby mine, which uses large gas-filled detectors to determine the direction of incoming dark matter. Observing the dark matter coming from the cosmos and producing dark matter in a particle accelerator (assuming that a particle is responsible for the dark matter) will combine to shed light on this mystery.

No one has ever seen evidence of proton decay, but most grand unified theories predict that proton decay should occur (though past experiments have indicated that the half-life of the proton is greater than 10^{32} years). The trick to observing proton decay is to have an exceedingly large volume of material in which the very rare decay products would be detectable. Possibilities include a large volume of water or liquid argon in which to detect radiation from such a decay. To go beyond the limits of past searches, these detectors would have to be hundreds of thousands of tons in size, and they would have to be deep underground to reduce background effects from particles coming from the sky. As noted before, a detector of this type also would detect neutrinos from space and could serve as the detector for neutrinos from a distant accelerator. The early proton decay experiments searched for decays of protons in water. Other detectors used different materials and more sophisticated tracing methods that are more sensitive to specific possible decay patterns of the proton.

Very energetic neutrinos (with energies well above the Terascale) might come from quasars, gamma ray bursts, black holes, or dark matter annihilation. Neutrino telescopes work by detecting light produced when such a neutrino encounters a nucleus. Some experiments look for this light in the ice of Antarctica, which is so clear that the light can travel for 100 meters undiminished. Other experiments look for it in the clear water of the sea or in a large lake. The first neutrino observatories—Amanda at the South Pole and the Baikal Neutrino Observatory at Russia's Lake Baikal—recently started operation. Others are now under construction or are being planned. The Antares experiment, deep in the Mediterranean near Marseilles, and Nestor, southwest of the Peloponnesian Islands at the deepest ocean site, will start operation in 2007. Amanda's successor in Antarctica, IceCube, is under construction and will be completed in 2010. A clever design and engineering approach makes the detector very modular and it is, in fact, already collecting data, with additional modules to be installed over the next few years.

Together with the research community, NSF has initiated a process to consider constructing a multidisciplinary deep underground science and engineering laboratory (DUSEL) and has selected two possible sites: the Homestake mine in South Dakota and the Henderson mine in Colorado. A decision on where to construct such a facility will be made later in 2006. DUSEL will offer an overburden of more than 6,000 meters-of-water-equivalent (that is, it will offer protection from cosmic rays equivalent to being under 6 km of ocean) and will have an initial suite of experiments that could include biological observations, dark matter experiments, a double beta decay experiment, and searches for solar neutrinos. In addition, the laboratory might contain a large cavern that would be suitable for a proton decay experiment.

Other experiments look for photons from the galaxy or beyond with energies of up to 1 TeV (see Box 3-6). These gamma rays might produce information about astrophysical accelerators, such as active galactic nuclei, pulsars, and supernova remnants, or perhaps about the origin of gamma-ray bursts. High-energy gamma rays might be produced when pairs of dark matter particles annihilate into pairs of photons. These would originate from the center of galaxies, where dark matter is most concentrated. Earth-based Cerenkov telescopes such as the Very Energetic Radiation Imaging Telescope Array System (VERITAS) and satellite experiments such as the Gamma-ray Large Area Space Telescope (GLAST) will search for these events.

Some evidence for ultra-high-energy cosmic rays (energies greater than 500 times the Terascale) has been reported, which conflicts with theoretical predictions that the gamma rays should have been slowed by their interactions with the cosmic background radiation. Using detector arrays roughly the size of Rhode Island, physicists at the High Resolution Fly's Eye (HiRes) experiment in Utah and the Pierre-Auger Observatory in Argentina are exploring this high-energy regime and trying to identify the sources of the highest energy cosmic rays. An outpost of the Auger experiment in the northern hemisphere could further help to pinpoint the sources of high-energy cosmic rays.

Telescopes can produce information about both dark matter and dark energy. They can look for distortions of the light from galaxies caused by the gravitational field of dense clumps of dark matter lying between a galaxy and Earth. They also can use these distortions, as well as supernovae, improved measurements of the cosmic background radiation, and the spatial distribution of clusters of galaxies, to learn about dark energy. Some proposed telescopes are ground based (such as the Dark Energy Survey, the Large Synoptic Survey Telescope, and the Panoramic Survey Telescope and Rapid Response System (PanSTARRS), while others will be launched into space so that they can cover more of the sky and look at more distant galaxies (an example is the Joint Dark Energy Mission).

Other telescopes are tuned to look at the CMB that remains from the moment

380,000 years after the big bang, when free electrons and protons cooled into hydrogen atoms. For instance, the European-led satellite Planck is scheduled to launch in 2007 and will analyze, with the highest accuracy ever achieved, the distribution and structure of the CMB. Other missions, such as NASA's Inflationary Probe, will search for the imprint of gravitational waves on the polarization of the CMB, a critical test that can distinguish among competing cosmological models.

INTERNATIONAL COOPERATION

International cooperation and collaboration have been prominent in particle physics since the field's inception in the first part of the 20th century. Scientists and laboratories around the world have engaged in both cooperation and healthy competition as the field has advanced. European, Asian, and other scientists from abroad have participated in experiments in the United States, and U.S. researchers have participated in efforts abroad.

Global Activity in Particle Physics

Some examples of the formal mechanisms the particle physics community has used to carry out international collaborations of various kinds are listed in Box 3-7. It should be noted that many successful international collaborations of the past decades began with grass-roots activities of interested scientists who then worked to obtain recognition by governments or government-to-government agreements.

As facilities on the scientific frontier have become more expensive to build and to operate, physicists from other countries have been asked to contribute financially to projects in host countries. The largest and most recent examples in the United States are the B factory experiment at SLAC and the CDF and D0 experiments at Fermilab. Roughly half of the collaborators on these experiments are from outside the United States, and the experiments are supported in part with significant financial contributions from abroad. Most of the international contributions to accelerator facilities have been in the form of scientific expertise and in-kind contributions to the detectors, as opposed to the building and operation of the accelerator.

Due to similar constraints for instrumentation and facilities in the global astronomy community, the United States has had great success partnering with other countries to construct and operate world-leading observatories such as the International Gamma-Ray Astrophysics Laboratory, the XMM-Newton X-ray Telescope, and the optical/infrared Gemini Observatory. Future observatories will include the Atacama Large Millimeter Array and GLAST.

BOX 3-7
Existing Mechanisms to Promote International Cooperation

Over the years the particle physics community has used a number of mechanisms for international discussion and planning. Current fora include:

- *The International Union of Pure and Applied Physics (IUPAP)*. This organization, chartered in 1933, is a member of the International Council for Science (ICSU, formerly known as the International Council for Scientific Unions). IUPAP is a nongovernmental union whose mission is to coordinate international activity in physics. It works through subject-area commissions and standing working groups or committees that are tasked with international coordination for more specific areas of physics.
- *The International Committee on Future Accelerators (ICFA)*. This working group of IUPAP was established in 1976 to facilitate international collaboration in the construction and use of accelerators for high-energy physics. It has taken an active role in developing plans for the ILC.
- *The Particle and Nuclear Astrophysics and Gravitation International Committee (PA-NAGIC)*. Created by IUPAP in 1999, this working group is charged with the coordination of non-accelerator-based international projects. PANAGIC has established two subpanels relevant to particle physics, one on high-energy neutrino astrophysics and one on gravity waves.
- *The Global Science Forum (GSF)*. Created by the Organisation for Economic Co-operation and Development (OECD), GSF is an organization of senior science policy officials from member countries. They meet twice yearly and discuss large science projects, including those in particle physics. GSF created a special group, the Consultative Group on High Energy Physics, which issued a report in June 2002 that contained a roadmap for high-energy physics extending to beyond 2020. Issues highlighted in the report include the legal structures, financial arrangements, governance, and roles of the host nations and laboratories for accelerator facilities.
- *The Funding Agencies for the Linear Collider (FALC)*. An informal group formed in 2003, FALC brings together representatives of the principal governmental agencies that fund research programs in particle physics. U.S. representation to FALC includes the NSF and DOE's Office of Science.

Perhaps the most important international collaboration in particle physics is the CERN laboratory in Geneva, which is a long-term cooperative effort of many European countries.[6] The construction programs for the detectors at the LHC,

---

[6]CERN member states are Austria, Belgium, Bulgaria, the Czech Republic, Denmark, Finland, France, Germany, Greece, Hungary, Italy, The Netherlands, Norway, Poland, Portugal, the Slovak Republic, Spain, Sweden, Switzerland, and the United Kingdom. Member states make a contribution to the capital and operating costs of CERN programs and are represented in the CERN Council, which is responsible for all important decisions about the organization and its activities. The United States is not a member but is granted observer status. Observer status allows nonmember states to attend Council meetings and to receive Council documents without taking part in the decision-making procedures.

along with the accelerator itself, also are examples of successes in international collaboration, with the United States and other non-CERN members contributing both financial and intellectual resources. The significant U.S. participation in the LHC project exemplifies some of the elements of a new era of global programs in particle physics.

During discussions about the high cost of excavating the tunnel for the Large Electron Positron Collider (LEP) at CERN, European researchers chose to examine possible future-generation accelerators to replace LEP at the same site. In 1985 the CERN Long-Range Planning Committee recommended installing a multi-TeV facility in the LEP tunnel after the completion of that program. In late 1991, the CERN Council agreed in a unanimous decision that the LHC was "the right machine for the further significant advance in the field of high energy physics research and for the future of CERN."[7]

When Congress terminated the construction of the Superconducting Super Collider (SSC) in 1993, the particle physics community and DOE recognized that the best practical opportunity to explore the Terascale within the next 10 to 20 years would be at the CERN-based LHC. At the request of DOE, HEPAP convened a panel to develop a new long-range plan for U.S. particle physics. It recommended that the United States participate in both the LHC experimental program and the construction of the LHC accelerator through significant contributions of in-kind components and cash for purchases of critical items in the United States. The particle physics community, DOE, and NSF strongly supported these recommendations. In early 1996, CERN's director general led a delegation to Washington to begin negotiations concerning a U.S. role in the LHC project. Around that time, CERN reached agreements for contributions to the LHC from Japan, India, Russia, and Canada, and NSF began to fund some LHC-related activities. The administration requested funds for strong U.S. participation in the LHC in its FY1997 budget; Congress then appropriated funds for both NSF and DOE to provide the U.S. contributions to the LHC. A very important step in this process was taken when Congress authorized DOE to enter into a formal agreement with CERN on behalf of the United States. U.S. officials signed the agreement with CERN in December 1997, promising to contribute $531 million to the LHC project over about 10 years. That investment is now nearly complete. This process of national initiative followed by international negotiation and agreement (resulting in a significant multiyear commitment from the United States) to invest in a facility abroad was an important achievement for the U.S. particle physics program and the U.S. government.

---

[7]CERN Press Release, PR12.93, December 17, 1993.

The CMS and ATLAS detectors being built for the LHC each have around 2,000 collaborators from all regions of the world. U.S. researchers make up about 20-25 percent of each detector collaboration, and the number of U.S. researchers is growing. By 2007, more than half of all U.S. experimental particle physicists are expected to be working at the LHC. The overwhelming reason for this shift is the planned conclusion of the U.S.-based experiments at SLAC, Cornell, and Fermilab. Many of the scientists in the university groups and laboratories that participated in the research program of these experiments are now transferring their efforts to the LHC. The model used by particle physicists to fund, build, and perform science with particle detectors has been and continues to be successful even at the largest scales.

Among recent projects, the J-PARC multiprogram accelerator complex was approved by the government of Japan, including an accelerator neutrino experiment, after which international involvement was welcomed. Significant non-Japanese funds (80 percent) have been raised to pay for one of the detectors at the facility. In general, if the science is exciting, scientists from around the world will want to join those efforts and will raise modest funds to participate. The director of KEK has said that if it is approved by the government, the new proton decay experiment HyperK will require international funds to move planning forward.

Accelerators around the world have thus far been built based on decisions made by a single country or laboratory; the exception has been the largest projects at CERN, such as the LHC (the CERN Council includes scientific and government representatives from each of the member states). The SLAC B factory accelerator was a U.S. presidential initiative, Fermilab's Tevatron was a U.S. decision, and constructing the SSC was a U.S. decision by President Reagan. The largest accelerator project to be successfully completed in the United States, the Spallation Neutron Source at Oak Ridge National Laboratory (with a cost exceeding $1.4 billion), was an internal U.S. decision. As is customary with DOE accelerator-based facilities, access will be open to scientists from around the world. DESY used a different model for the HERA accelerator: The plan was to build components of the accelerator in several countries as in-kind contributions to be assembled at the main facility. Although DESY had hoped for substantial contributions, the final non-German fraction was 15 percent. Even the Euro-XFEL, a $1 billion project just under way and being hosted at DESY, was approved by the German government, after which contributions amounting to 50 percent of the cost were sought from Europe. This approach appears to have been successful because of the strong support from the user community for this facility.

Europe, through CERN, recently took the next step in formalizing its regional planning activities. A group has been established through the initiative of the CERN Council to develop a strategy that addresses the main thrusts of particle physics in Europe, both accelerator-based and non-accelerator-based, including R&D for novel accelerator and detector technologies. The strategy is designed to address collaboration between the European laboratories, coordinated European participation in world projects, the visibility of the field, and knowledge transfer beyond the field. Since CERN is an international organization, its Council is composed of government representatives. Thus, approval by the CERN Council invokes the treaty relationship between each government and the CERN organization, creating a binding agreement among the individual governments.

The opportunities for international collaboration in particle physics and the challenges posed have never been greater. More rigorous international prioritization of new particle physics research opportunities and greater leveraging of international funding could have great benefits as particles physicists seek to answer the exciting questions now before the field. Such benefits, however, can only be realized through genuine cooperation both among scientists and among the government agencies sponsoring their work. The most extensive current example of international collaboration is the set of activities that surround the planning and R&D phases for the proposed ILC.

The International Linear Collider

Particle physics research communities around the world have declared that the ILC is the highest priority project after the LHC.[8] The ILC promises to provide answers to a host of the most important questions in particle physics. It is clearly of a scale where decisions on design, funding, and operation must be international from the start. (See Appendix A for additional analysis of the path forward.)

The committee felt strongly that, if possible, the ILC should be located near an existing particle physics laboratory to take advantage of existing resources and talent.[9] Past experience with the SSC, as well as current experience with the LHC, shows the advantages of undertaking new projects with existing facilities and talent. As the only laboratory devoted primarily to particle physics, Fermilab is an obvious candidate site. It is attractive as a potential site for the ILC because of its existing laboratory and physical plant infrastructure. Like CERN in Europe, Fermilab has a critical nucleus of accelerator expertise that could play a significant role in the ILC. Fermilab has successfully built, operated, and upgraded the

---

[8]Among 28 large-scale facilities across all of the physical sciences, DOE's Office of Science deemed U.S. participation in the ILC the highest priority initiative for the mid-term planning horizon.

[9]This sense is supported by a number of other reports considering possible site selection criteria for the ILC.

Tevatron, one of the most sophisticated accelerators in the world. In collaboration with DESY in Germany and other laboratories, Fermilab also has developed expertise with superconducting radiofrequency technology, the choice for the ILC. Fermilab must provide the leadership necessary to mobilize a coalition of U.S.-based resources and facilitate U.S. participation in the ILC.

The ILC has been an international effort from its inception and should continue to be pursued as a global venture. In 2005 the U.S. effort in ILC R&D was budgeted for $25 million; other regions of the world have invested much more. For instance, European governments invested more than $50 million in 2005. Integrated over about 5 years, the Japanese and European investments in ILC R&D total at least several hundred million dollars.

A critical element of any U.S. strategy to move forward with the ILC beyond the initial R&D phase coordinated by the Global Design Effort (GDE) will be the formation of an entity capable of negotiating both scientific and financial matters with the other expected regional partners. At present, the association between the U.S. program (through DOE and NSF) and the GDE is only informal (for more on the GDE see Appendix A). Moving forward on the ILC will demand new mechanisms of cooperation and agreement among the research agencies of many nations. Several such agencies have begun to discuss the ILC project at an international level through the FALC group, an informal body composed of representatives from relevant funding agencies from Canada, France, Germany, Italy, Japan, the United Kingdom, the United States, and CERN. Formed in 2003, FALC provides a forum to discuss funding issues, policy strategies, and progress toward designing an ILC. As this effort moves forward, the decision-making process will be complex and will require simultaneous discussions at the scientific level and at various governmental levels that transcend the FALC group. Experience with other international joint ventures (such as ITER and the LHC) demonstrates the potential for success in sophisticated international agreements of this kind.

A PATH FORWARD

Over the next 15 years, today's international collaboration, already extensive, will need to intensify to effectively address the challenges on the scientific frontier. The committee believes that particle physics should evolve into a truly global collaboration that allows the particle physics community to leverage its resources, prevent duplication of effort, and provide additional opportunities for particle physicists throughout the world.

This prioritization process could lead to a new model for international collaboration in particle physics. For example, each country or region could specialize to some extent in programs sited in their country or region and then play a relatively smaller role in programs sited elsewhere. Such an evolution would be in keeping with the framework proposed in *Allocating Federal Funds for Science and Technology*.[10] Among the report's recommendations, two are particularly noteworthy:

> The President and Congress should ensure that the [federal science and technology] budget is sufficient to allow the United States to achieve preeminence in a select number of fields and to perform at a world-class level in the other major fields.
>
> The United States should pursue international cooperation to share costs, to tap into the world's best science and technology, and to meet national goals.

Both goals would be met if the United States were to participate in a worldwide effort to plan particle physics research from a global perspective. Furthermore, the ILC could serve as the model for a global program, since the early planning has already started from a global perspective rather than from the perspective of an individual country. This planning process could ultimately be expanded into many other areas of particle physics. While meeting these goals would serve the interests of particle physics and fulfill the public policy objective of using resources in the most efficient manner, success can only be achieved through multilateral agreements between governments and/or government agencies, not unilaterally. This is a challenging task but one that must be done given the environment the committee believes will evolve over the next 15 years.

The tools of particle physics have evolved significantly over the past 50 years. Originally particle physics was a small field; individual scientists could construct particle accelerators (first tabletop and then room-sized cyclotrons) and detectors (plates of film) in their own laboratories. As the science drove accelerators to higher energies, the scale of projects continued to expand. In the modern era, the most recently designed and constructed machines require literally hundreds of scientists and engineers. Partly because of the demands for high performance and partly because of the eclectic nature of the investigations, particle physics projects in the United States are constructed (and then operated) with sizable involvement of scientists and engineers, more so than in some other fields, such as magnetic fusion.

The planning process for particle physics in the United States historically has

---

[10]NRC, *Allocating Federal Funds for Science and Technology*, Washington, D.C.: The National Academies Press, 1995, pp. 14, 16.

involved more than one government agency. Broad involvement of the particle physics community has been achieved by creating a variety of advisory committees, such as HEPAP and its subcommittees, which advise DOE and NSF; program advisory committees at the major laboratories; and National Research Council committees that periodically review the field from a broader perspective.

Nearly all of the larger national laboratories have had an important program in particle physics, which is a tribute to the broad appeal and importance of particle physics to the physical sciences. Argonne National Laboratory, Brookhaven National Laboratory, Cornell Laboratory for Elementary Particle Physics, Fermilab, Lawrence Berkeley National Laboratory, Lawrence Livermore National Laboratory, Los Alamos National Laboratory, SLAC, Thomas Jefferson National Accelerator Facility, and others have all contributed to scientific and technological advances in particle physics. Different laboratories have pursued different initiatives by developing machines capable of new investigations, whether involving higher energies, higher intensities, or beams of different particles. A strong and healthy national program was maintained through intense but healthy competition (for both resources and personnel) among the variety of different projects. This situation is changing, however. As Fermilab becomes the only laboratory devoted entirely to particle physics, the system of planning and coordinating efforts will have to evolve as well.

Approved projects are subject to ongoing external reviews by experts from the broader community, at least during their construction phase. Large projects that overlap the interests of more than one agency need a planning and review process that is effective and not duplicative. New initiatives need to be able to bubble up within the field, but large-scale efforts need a coordinated decision process to establish their overall priority, a process that is national rather than based on a single laboratory or government agency. The astrophysics community has achieved this goal with a structured decadal review process. In particle physics, HEPAP has been the leading source of advice to the U.S. government, and its recently established P5 subcommittee offers program review and coordination at a higher level than the laboratory program committees for larger ventures, although this mechanism is new and has not yet been effectively deployed. The advisory apparatus has been evolving, and the emerging structure of tactical subfield-specific scientific assessment panels (such as in neutrino physics or dark energy) feeding into P5 and HEPAP for the formulation of strategic guidance is a step in the right direction. The challenge for federal agencies is to continue to get the needed community input but to avoid creating an overlapping and possibly contradictory set of advisory groups and panels. This requires some interagency coordination and works best when there is a stable, long-term planning process that the community understands and accepts as authoritative.

No description of developments in particle physics would be complete without acknowledging that, as in any area of science, not all experiments have achieved their goals. Some experimental disappointments inevitably accompany the exploration of the unknown and are a part of the process responsible for scientific progress. Other experiments failed to find what they were looking for but instead found other very important results.

Within the U.S. program, the biggest disappointment was the collapse, in the early 1990s, of the SSC program. This accelerator was designed to access extremely high energies, substantially higher even than the energies that will be reached at the LHC when it begins operation. The cancellation of the SSC was a severe blow to U.S. scientific leadership and to progress in particle physics (see Box 3-8).[11] A number of lessons were learned from this difficult and costly experience about effective ways to proceed with large scientific projects involving international partnerships.

First, effective international partnerships require the meaningful participation of all parties from planning and design through conduct of the experiments. Second, very detailed design parameters are essential before starting construction and before announcing any cost estimates. Third, effective simulation models are needed (and have since been developed) to help provide more reliable and robust cost estimates and performance expectations. Fourth, effective, integrated management that takes advantage of existing resources and infrastructure is critically important. These hard-won lessons are being implemented in studies surrounding the proposed ILC.

OPPORTUNITIES AHEAD

The different tools of particle physicists—high-energy accelerators, intense particle beams, and ground- and space-based observations of the universe—will all be necessary to take the next steps in answering the fundamental questions of particle physics. New physics at the Terascale will be revealed and studied at the LHC and the ILC. Neutrino beams can yield further insights into the properties of many other particles. And a full understanding of dark matter and dark energy will require the tools of particle astrophysics.

---

[11]At the same time, Europe, through CERN, was able to move ahead with a set of objectives articulated (informally) much earlier. The usual pattern is that new accelerators stand on the shoulders of their predecessors. At the time of construction of the underground tunnels for CERN's LEP in the early 1980s, then director general John Adams had a vision for a natural progression from LEP to an advanced proton collider in the same tunnel, such as the LHC, that would make use of the existing infrastructure.

BOX 3-8
A Brief History of the Superconducting Super Collider

The idea for a colliding proton-proton accelerator with energy of 20 TeV per beam was first discussed at a series of workshops held in 1978 and 1979 by ICFA. Plans for the collider were discussed extensively at a summer study sponsored by the American Physical Society in 1982 in Snowmass, Colorado. Even then, the project was recognized as a multi-billion-dollar undertaking that would require substantial international collaboration. In 1983, after several subsequent workshops, HEPAP recommended that DOE seek "immediate initiation of a multi-TeV high-luminosity proton-proton collider."[1]

In 1984 DOE approved the establishment of the Central Design Group for the SSC under the management of the University Research Association (URA), a consortium of universities that also manages Fermilab. By 1986 the design group, based largely at the Lawrence Berkeley National Laboratory campus, had produced a conceptual design with a price tag of more than $4 billion. DOE recommended moving forward with the project, and in January 1987 the Reagan administration made the project a national initiative. The selection of a site near the town of Waxahachie, Texas, was announced on November 10, 1988.

Even at that point, several of the tensions that would later become critical factors in the cancellation of the project were apparent. Proposals for the SSC from the administration posited significant financial contributions from other countries. But parts of the administration and several powerful senators saw the SSC as primarily a U.S. undertaking designed to reestablish national supremacy in high-energy physics. As a result, international collaboration was not part of the project from the beginning and was pursued only after Congress had already committed to the project.

The management of the project also was becoming controversial. DOE officials had doubts that physicists could manage a project the size of the SSC. Responding to these doubts, the URA's proposal to build the SSC featured partnerships with industrial firms that had experience in managing large construction projects. This unusual management scheme contributed to dissatisfaction among the members of the Central Design Group, many of whom declined to continue working on the project.

Increases in the estimated cost of the SSC were another source of concern. After the selection of the Texas site, DOE submitted a revised cost estimate to Congress of $5.9 billion in early 1989. However, work was under way at that time to incorporate into the design several additional features felt to be necessary, such as a more powerful proton injector ring and better superconducting dipole magnets. These and other modifications added more than $2 billion to the cost, yielding a revised estimate of $8.25 billion in February 1991.

In key votes in 1989, 1990, and 1991, both the House and Senate supported the SSC. But misgivings about the project were growing. The Europeans were working on plans to build their own proton-proton collider at CERN. The Japanese reportedly were willing to contribute to the construction costs of the collider, but they wanted a personal request from either President Bush or newly elected President Clinton, which, for various reasons, never came. The project also was being criticized by other scientists, including some physicists, who saw its funding undermining support for other areas of research.

In June 1993 the House voted 280 to 150 to terminate the SSC project. The Senate continued to support the project and prevailed in conference to have funding included in the DOE appropriations bill. Then, on October 19, 1993, the House rejected the entire appropriations bill by a vote of 282 to 143. Support for the SSC subsequently collapsed. Congress directed that the $640 million appropriated for the project in 1994 be used to terminate the project. After

continued

BOX 3-8 Continued

expenditures of approximately $2 billion, the contracts for the superconducting magnets were canceled, the entrances to the 15 miles of tunnel already dug were blocked with rock, and the employees of the SSC laboratory began looking for new jobs.

Foreign assistance for the SSC had been expected to come not as cash that could be spent within the United States but as in-kind contributions—chiefly furnished materials and manufactured items such as superconducting magnets, cryogenic systems, computers, or other electronic components. Projected international cooperation did not materialize, which meant that the entire cost of the project would have to be borne by U.S. taxpayers. The proponents of the SSC had argued that many countries were eager to participate and contribute financially if only Congress would demonstrate good faith by funding the SSC more fully—a classic chicken-and-egg problem. By 1992, however, India was the only nation to pledge any support for the SSC project—a total of $50 million, or about half of 1 percent of the projected total cost. The European Community, which had been planning its own supercollider (which became the LHC), was not a realistic source of funding for a U.S. project, many contended. Japan had been expected to be a major contributor, but the Japanese government resisted pressures by the U.S. government to become one. Some contend that Japan, which may have been willing to commit up to $1 billion, was reluctant to proceed until more formal government-to-government agreements to provide a framework for cooperation were worked out.

According to an editorial in *Science,* "In its quest for big bucks for the particle accelerator, the United States appears to have ignored the golden rule for getting major contributions from Japan: links must be built at ground level before an official approach for funds."[2]

The cancellation of the SSC not only was a severe blow to the U.S. program in particle physics and U.S. scientific leadership, but it also delayed progress in particle physics by postponing direct exploration of the Terascale with a proton collider. The LHC now being built at CERN shares many of the scientific goals of the original SSC, but it has a higher particle intensity and a lower energy. The proposed ILC would differ significantly from both the SSC and the LHC by employing colliding electrons to probe the Terascale. The ILC proposes using a technical approach and management structure entirely different from that for the SSC (see details in the text).

[1] HEPAP Report to the Department of Energy and the National Science Foundation, 1983.
[2] *Science,* April 5, 1991, p. 25.

The strong attraction of Terascale physics is underscored by the convergence of interests from distinct scientific areas. From cosmology, there is growing interest in dark matter and dark energy. From particle physics, there is great interest in supersymmetry, in the origins of mass, and in Einstein's dream that all the forces can be unified. This convergence is what makes the Terascale so compelling. The intersection of scientific interests is often a signal that major new discoveries are on the horizon. Thus, the committee feels that explorations of the Terascale have enormous scientific potential.

Addressing these scientific challenges can be part of a national commitment to renew the country's portfolio in basic research to "maintain the flow of new ideas that fuel the economy, provide security, and enhance the quality of life."[12] Moreover, it is a deeply human endeavor that involves some of the most world's most talented scientists, engineers, and students.

---

[12]NAS, NAE, and IOM, *Rising Above the Gathering Storm: Energizing and Employing America for a Brighter Economic Future*, Washington, D.C.: The National Academies Press, 2005 (Prepublication), p. 20.

4

The Strategic Framework

This chapter presents the committee's strategic framework for the U.S. particle physics program. This framework is based both on the unusually exciting scientific challenges now facing particle physics and on the committee's belief that a strong role in this area is necessary if the nation is to sustain its leadership in science and technology over the long term. The committee's chief findings and recommended action items, which appear in the next chapter, are based on the strategic framework and budget scenarios presented in this chapter.

THE SCIENTIFIC CHALLENGE

Elementary particle physics advances by posing deep questions about the origin and character of some of nature's most fundamental entities and conducting experiments to answer those questions. The experiments not only yield new knowledge of nature's laws and develop new technologies, they also almost inevitably lead to even more profound questions. On this voyage of discovery, major scientific breakthroughs are achieved when important questions begin to intersect in unexpected ways, producing a deeper and more fundamental understanding of the phenomena being studied. Elementary particle physics is at such a moment now, when great questions are before it and the field is poised to answer them.

A century of revolutionary discoveries, together with the development of new technologies, has produced a dazzling array of scientific challenges in particle physics. The scientific challenges and opportunities for discovery on both the scientific and technological frontiers, particularly for exploration at Terascale energies, are extraordinarily exciting. The opportunities now accessible to particle physics include moving beyond the limitations of the Standard Model, exploring further the unification of forces, probing the origin of mass, uncovering the dynamic nature of the vacuum, deepening the understanding of stellar and nuclear processes, and investigating the nature of dark energy and dark matter. These possibilities suggest that a great deal of new physics may be discovered in the next generation of experiments.

THE POSITION OF THE U.S. PROGRAM

Despite an extraordinary tradition of U.S. leadership in this area of science, the intellectual center of gravity in most areas of particle physics will move abroad with the termination in the next few years of the B-factory experiment at SLAC, the CLEO experiment at Cornell, and the CDF and D0 experiments at Fermilab. Moreover, this will occur just at the moment when especially exciting and important scientific opportunities have appeared on the horizon.

The U.S. program in elementary particle physics is therefore at a crossroads. On the one hand, there is an opportunity to reallocate substantial resources to begin exploiting new opportunities as existing experimental programs are completed over the next 2 to 4 years. Further, the United States has the necessary human capital, technology, and industrial expertise to be a leader in the pursuit of the scientific challenges of elementary particle physics. Indeed, it has a large pool of particle physicists, accelerator scientists, advanced students, and other talented researchers who can identify and pursue the most important and challenging questions in the field.

On the other hand, if the United States is to exploit these opportunities—and in the process fire the imagination and creativity of the next generation of students and scientists—decisive actions must be taken now. These actions will require a new strategic framework that establishes priorities designed to ensure a leadership role for the United States in the decades ahead and points to the difficult decisions required to act on those priorities. Moreover, regaining the long-term momentum of the program in elementary particle physics and reestablishing a position of leadership require a willingness to take scientific and technological risks and to consider important institutional transformations. While this effort will be a demanding one, the failure to take up the challenges might lead some of the best U.S. scientists and students to disperse abroad or to other fields of endeavor, undermining the nation's opportunity to continue to play a leadership role in this fundamental scientific area.

Fortunately, because several of the nation's most important experiments in particle physics are coming to an end, reallocation of resources within the program would allow the U.S. particle physics research community to begin to implement a strategic vision that is consistent with contemporary scientific developments and with sustained U.S. leadership in the field.

The committee feels strongly that because of the increasing cost and complexity of particle physics experiments, and the need to deploy public funds in the most effective and responsible manner, it is more important than ever for all the large programs in particle physics to leverage their resources by working together internationally. The community of particle physicists has a strong tradition in this area, but that tradition needs to be enhanced. There is an increasing need for particle physics programs in the United States and elsewhere to take fuller advantage of important experiments proceeding in other countries. Moreover, the key sponsors of national and multinational programs need to allow for the serious consideration of new and imaginative arrangements. Such arrangements would not only serve the cause of scientific progress, they might also be the only way to provide scientists and their students around the world the opportunity to address those areas of particle physics to which they can make the greatest scientific contribution. This type of transformation cannot be accomplished by a single country or region. It requires the mutual collaboration of all major partners. From the perspective of the U.S. program in particle physics, such arrangements could be of great value as they would give U.S.-based researchers better access both to a wider portion of the scientific frontier and to a wider range of opportunities. Such changes would strengthen the knowledge base of the entire U.S. scientific enterprise.

In crafting a strategic plan for the U.S. program in particle physics over the next 15 years, it is important to identify and balance the risks that are inherent in any such activity. First, there are scientific risks. Frontier experiments push the boundaries of human experience; it is never certain what lies beyond current knowledge. Because of that, the particular shape, focus, and character of the next set of experiments can be expected to evolve, at least in part, in unexpected and unpredictable ways. Second, there are unavoidable structural risks. Experimental facilities for elementary particle physics are constructed and supported by government funds, so planning must factor in uncertainties surrounding future government investments in science, in general, and in elementary particle physics, in particular. Third, there are special risks associated with the most important of the next generation of experiments, the proposed ILC. The ILC is a very large project, and important and critical investments must be made before it is certain that an international consortium can be assembled to construct and operate the collider and before a decision can be made about where it will be sited. A willingness to accept these risks is an inevitable aspect of a leadership position at the scientific and technological frontiers.

The committee emphasizes that while investment in specific scientific projects always carries risk, leadership in science is central to the nation's strategic and economic security. Science is concerned with the investigation of the unknown, so one cannot be certain in advance of the dividends that will be achieved, either in new scientific understanding or in novel technological developments. Any large scientific project carries additional risks because new experiments push technology to new frontiers. In this respect, however, elementary particle physicists have accumulated an enviable record in meeting technological challenges. In the process, they have provided society with an array of useful innovations in science, medicine, and industry (such as in computing and in medical imaging and treatment). **It should also be kept in mind that there are greater risks in not exploiting scientific opportunities and in forgoing the potential benefits to society and human development.** The risks of inaction are difficult to assess fully, but they may be quite significant.[1]

In crafting its recommendations, the committee first articulated a set of strategic principles designed to provide an overall framework for the U.S. program in particle physics. These principles are presented in the remainder of this chapter. Within the context of these strategic principles, the committee, on the basis of its specific findings, worked out a set of recommended action items reflecting the priorities that it believes ought to guide the program over the next 15 years. The findings and recommended action items are presented in Chapter 5.

THE STRATEGIC PRINCIPLES

In the modern era, leadership in particle physics does not mean dominance. Rather, it means playing a central role in new scientific discoveries, which can be done by taking initiatives on the scientific frontier, accepting risks, and catalyzing partnerships with colleagues at home and abroad. In the contemporary world of particle physics, none of the national and/or regional programs is—or can be expected to be—in an overall leadership position in the sense of command and

---

[1]A substantial body of literature exists on this topic, but for this discussion, consider the following comment from the *American Competitiveness Initiative*, p. 4, a publication of the U.S. Domestic Policy Council in February 2006: "Our prosperity is no accident. It is the product of risk-takers, innovators, and visionaries."

control or singular dominance. There are, however, a small number of national and regional programs that are currently exerting leadership in the sense of providing a continuing stream of significant intellectual and experimental contributions to the most important issues on the scientific frontier. As a result, these programs have a major influence on the evolving profile of the field and are in the best position to exploit the scientific and technological developments that emerge and to initiate and mobilize joint international efforts. Thus, the committee's practical definition of leadership (provided above) is a statement of aspiration in all these areas (intellectual relevance, active participation on the frontiers of science and technology, acceptance of risk, and catalysis of international partnerships), and it is the kind of leadership that the United States should seek to maintain in the years ahead. Therefore, in considering the U.S. program in particle physics for the next 15 years, the committee sought not only to pursue the most compelling scientific opportunities but also to reestablish a clear path to a position of leadership in particle physics.

Strategic Principle 1: The National Importance of Elementary Particle Physics. The committee affirms the intrinsic value of elementary particle physics as part of the broader scientific and technological enterprise and identifies it as a key priority within the physical sciences.

The current scientific and technological prowess of the United States is due in no small part to the nation's investments in basic research in the physical sciences. Elementary particle physics is an important part of this research portfolio, through both its contributions to a variety of scientific fields and its being an integral part of the broader inquiry into the basic workings of nature.

One example of the interplay between particle physics and other fields of physics is the development and application of a set of mathematical tools known as quantum field theory. Quantum field theory generalizes the principles of quantum mechanics to situations where the number of particles is not constant, and it provides an exhaustive framework for calculating complex phenomena. Quantum field theory has now become a general tool for a wide variety of theoretical physicists. For instance, condensed matter physicists use quantum field theories to describe phenomena such as superconductivity and phase transitions. In fact, certain advances in particle physics theory can be traced to inspirations from condensed matter uses of quantum field theory. Other examples of the intellectual connection between particle physics and the broader enterprise of physics involve the joint development and deployment of scientific instrumentation. For instance, it was the advent of large-scale semiconductor manufacturing in the 1980s that led to the development of a new generation of particle detectors using large surfaces of purified silicon; later, the technologies perfected by particle physicists found application in space-based observing platforms such as the GLAST satellite. Vice versa, particle physicists' interest in novel radiation-hard particle detection technologies led to the development of thin-film diamond sensors, which now have emerging applications in medical diagnostics and monitoring.

Perhaps the most important contributions of particle physics to the broader economy in recent years came from the development of the key protocols that underpin the World Wide Web at the CERN laboratory. SLAC was the first U.S. entity connected to the Web; Fermilab was the second. Building on the backbone of the already existing Internet, this new way of sharing information has revolutionized the way the world communicates and does business. The synergy between particle physics and cyberinfrastructure has played a strong role in the history of both fields. As new computing, information sharing, and data handling capabilities have become available, particle physics has embraced them and has been instrumental in developing many of the advances. Even today, physicists are working with their colleagues in computer and information science to implement architectures for shared computing access to the LHC experiment.

These advances arose because of a synergy between particle physics and other developments in science and technology; that is, the committee does not claim that particle physics is the best or only driver of such innovations. Rather, it argues that a strong program in particle physics is an essential element of an overall strategy to foster such breakthroughs.

Most important, as described in Chapter 3, the committee identifies elementary particle physics as a research effort that is poised to make transformative discoveries in the immediate future. The frontiers of human understanding are always advancing, yet the committee was struck by the tremendous discovery potential of particle physics over the next decade. Furthermore, the emerging connections among particle physics, astrophysics, cosmology, and nuclear physics are extremely promising signs of breakthrough opportunities.

There is every expectation that discoveries at the Terascale will ripple across the fields of science as new insight is gained into the nature of space and time, energy and matter.

Strategic Principle 2: U.S. Leadership. The U.S. program in elementary particle physics should be characterized by a commitment to leadership within the global particle physics enterprise.

The argument for a leadership role is multidimensional. First, the committee believes that leadership in this important and challenging area of science is critical to the overall strength of U.S. science and its role as an engine of economic growth through innovation. The connection between economic leadership and the physical sciences and mathematics has been strongly articulated in the recent National

Academies report *Rising Above the Gathering Storm*.[2] Particle physics is critical in this respect. It has a strong position at the forefront of technology, and its quest to understand elementary particles and fundamental phenomena connects it to many other areas of science as well as to industry. This is particularly true for accelerator R&D, which has created the accelerators that generate radiation for medical therapies and the x-ray beams that are now pushing the edge of bioscience and materials science. It has also been true of the other parts of particle physics in a variety of ways.

Second, unless the United States undertakes the challenge of leadership, scientists here will be unable to work effectively with their colleagues abroad. As asserted in *Allocating Federal Funds for Science and Technology*, "Science is a global enterprise in which the United States must participate, for its own benefit and for that of the world."[3] However, owing to the increasing capabilities of particle physics research programs in other countries, as well as the increasing cost of experiments, it is neither desirable nor feasible for the United States, or any other country, to host experimental facilities in every area of elementary particle physics. Instead, the United States must become a leader in particle physics through a combination of efforts: investing strategically in projects located in other countries, hosting particle physics projects with some of the greater potential for discovery, and ensuring that U.S. programs make the best use of particle physics personnel, facilities, and resources. That is, to remain competitive, the United States must seek collaborations that confer mutual advantage.

Third, occupying a leadership position will ensure that the United States reaps the dividends of new discoveries and ensures vitality for its next generation of scientists. As described in *Globalization of Materials Research and Development: Time for a National Strategy*,[4] ensuring U.S. access to cutting-edge science and technology, no matter where the next breakthroughs may occur, is a key reason for staying active at the frontiers of research. Permanently abandoning leadership in particle physics will have profound consequences. Not only will U.S. scientists, students, and engineers fall behind their colleagues in the rest of the world, but our nation will have given up on one of the key drivers of scientific and technological innovation.

With respect to future international joint efforts that might be based outside

---

[2]NAS, NAE, IOM, *Rising Above the Gathering Storm: Energizing and Employing America for a Brighter Economic Future*, Washington, D.C.: The National Academies Press, 2005 (Prepublication).

[3]NRC, *Allocating Federal Funds for Science and Technology*, Washington, D.C.: National Academy Press, 1995, p. 16.

[4]NRC, *Globalization of Materials Research and Development: Time for a National Strategy*, Washington, D.C.: The National Academies Press, 2005, p. 2.

the United States, the committee identified some neutrino physics experiments, a proton decay experiment, and/or a super-B factory as important examples. As already noted, the proposed ILC will certainly require an international joint effort; it also will require a major commitment of U.S. particle physics resources to succeed wherever it is based. The United States is already an active participant at the LHC at CERN as well as at other laboratories abroad, such as DESY in Germany, KEK in Japan, and the Sudbury Neutrino Observatory (SNO) laboratory in Canada. The U.S. particle physics program should continue to seek international partners to share the costs of U.S.-based efforts, just as the United States invests in overseas efforts.

> **Strategic Principle 3: A Global Particle Physics Program. As the global particle physics research program becomes increasingly integrated, the U.S. program in particle physics should be planned and executed with greater emphasis on strategic international partnerships. The United States should lead in mobilizing the interests of international partners to jointly plan, site, and sponsor the most effective and most important experimental facilities.**

The next generation of experiments will require more complex and expensive experimental facilities, including underground laboratory spaces for neutrino physics, dark matter searches, and proton decay experiments; possible upgrades of the LHC accelerator and detectors; intense neutrino beams and associated detectors for a second generation of long-baseline neutrino oscillation experiments; ground- and space-based efforts for particle astrophysics experiments; a possible future super-B factory; and, most ambitious of all, the ILC. One testament to the success of the pooling of international (predominantly European) resources and talents is the CERN laboratory in Geneva.

To achieve and maintain a leadership position in the global particle physics program and to maximize the return on the public resources invested, the United States must take the initiative in establishing joint programs aimed at exploiting the scientific potential of the largest, most complex, and most expensive of the next generation of experimental facilities. This implies that the United States should be willing to provide and be the lead investor in the appropriate facilities for some major part of the science at the forefront of the field and to welcome scientists from abroad as partners. It also implies that the United States should be willing to invest in important scientific opportunities or key experimental facilities located abroad. A critical (and often overlooked) aspect of participation in such a global program is the need for international discussion and coordination from start to finish of a project; that is, nations should consider and consult potential partners for a candidate project before, during, and long after the design, development, and engineering stages of a project begin. Just as in business relationships, valued and productive partnerships spring from early joint ownership of the project.

The strategic objective is to work with the nation's scientific partners abroad and their sponsors to forge an international partnership that deploys public investments in the most efficient and effective manner.[5] The committee believes that the globalization of scientific research, especially in particle physics, has opened a new path to leadership. For the United States to be globally competitive, attain national goals, and realize the most compelling scientific opportunities, the nation must plan and pursue the most critical ventures with international partners.

Strategic Principle 4: The Necessary Characteristics of a Leadership Program. The committee believes that the U.S. program in elementary particle physics must be characterized by the following to achieve and sustain a leadership position. Together, these characteristics provide for a program in particle physics that will be lasting and continuously beneficial:

- **A long-term vision,**
- **A clear set of priorities,**
- **A willingness to take scientific risks where justified by the potential for major advances,**
- **A determination to seek mutually advantageous joint ventures with colleagues abroad,**
- **A considerable degree of flexibility and resiliency,**
- **A budget consistent with an aspiration for leadership, and**
- **As robust and diversified a portfolio of research efforts as investment levels permit.**

The last of these characteristics deserves special emphasis. A broad array of scientific opportunities exists in elementary particle physics, and it is not possible

---

[5]This emergent strategy is not unique to particle physics. As Lynn and Salzman note, there are "strong possibilities that the nation can benefit by developing 'mutual gain' policies. Doing so requires a fundamental change in global strategy. The United States should move away from an almost certainly futile attempt to maintain dominance and toward an approach in which leadership comes from developing and brokering mutual gains among equal partners" (L. Lynn and H. Salzman, "Collaborative Advantage," *Issues in Science and Technology*, Winter 2006, p. 76). They term this approach "collaborative advantage" and say, "It comes not from self-sufficiency or maintaining a monopoly but from being a valued collaborator at various levels in the international system."

to foretell which of them will yield important new results soonest. Particle physics, like all other elements of the scientific enterprise, explores the unknown, and this inevitably requires shouldering some uncertainty. Thus, it is important to maintain a diverse and comprehensive portfolio of research activities—from theory to accelerator R&D to the construction of new experimental facilities to efforts to probe entirely new areas. Two of the greatest discoveries of the last decade—the discovery of nonzero neutrino masses and of dark energy—were completely unexpected, underscoring the need for a variety of approaches to current scientific challenges.

Even during a period of budgetary stringency, sufficient funding and diversity must be retained in the pipeline of projects so that the United States is positioned to participate in the most exciting science wherever it occurs. It is essential, therefore, to follow a mixed strategy that encompasses a variety of experimental approaches, arrangements that allow for the most advanced training of the next generation of scientists, investments in future detector and accelerator technologies, adequate computational resources, support for theoretical work, and the capacity to support small and innovative experiments. The relatively flat funding of the U.S. particle physics enterprise over the past decade has, unfortunately, forced a relative reduction in its diversity. Moreover, uncertainties over future support make some investigators more conservative in their research, leading them to work on more established, predictable topics.

For full participation in the international arena, the United States must coordinate, and in some cases subordinate, its planning to international planning and advisory structures such as IUPAP and ICFA. It is important to design mechanisms whereby joint programs incorporate the best ideas from all around the world. This means that duplicative preliminary work on projects must be supported for the best possible approaches to emerge. The effort to eliminate duplication of large projects should not end up suppressing the development of competing approaches too early in the process. At the same time, some international mechanism is needed to ensure that only the most promising approaches are supported.

The breadth of the U.S.-based program is an important factor. The U.S. particle physics program has benefited from a strong tradition of investments in the human, institutional, and physical infrastructure. For instance, the United States has been at the forefront of advancing the theoretical underpinnings of particle physics, which has had a profound effect on the shape of the experimental program. In turn, new theories have emerged from experimental results. The close relationship between theory and experiment has been a key driver of U.S. leadership in this field, and it is important to nurture this relationship. Another example of this tradition is the historical stewardship of accelerator science and technology by the nation's elementary particle physics program. Particle accelerators continue to affect a broad spectrum of scientific and technological activities. Advanced research into new accelerator technologies is vital to the future of accelerator-based elementary particle physics as well as to emerging technologies in other areas. The United States should strive to remain a lead player in this area.

The success and vitality of the scientific enterprise depend on a distinctive set of institutional arrangements for training new scientists. The committee views the current role of university-based students, postdoctoral researchers, and faculty as a critical component of the particle physics enterprise that strengthens national capabilities in both education and science. The strength of the university-based program also depends directly on a healthy, competitive peer-review system that identifies and supports the best science.[6] The framework of competitive peer review ought to govern the allocation of resources to the greatest extent practicable. Fair competition among competing ideas, be it at the individual investigator level or at the level of laboratory program initiatives, helps select and support the most compelling, ripest for exploitation of the science.

THE BUDGETARY FRAMEWORK

Recent Trends in Support for the U.S. Particle Physics Program

The U.S. program in elementary particle physics has not experienced any real growth in a decade. The committee estimates that over the last 5 years (FY2001 through FY2006) funding for this area of science declined by 5 percent in real terms (see Box 1-4). Some of the key U.S.-based experimental facilities in elementary particle physics are either being converted to serve other uses (the SLAC linear accelerator and the CESR accelerator) or are coming to the end of their scientific lives (the CDF and D0 experiments at Fermilab's Tevatron). This provides an opportunity to strategically reallocate these funds as part of a new and exciting long-term vision for the U.S. program, which, despite the circumstances, may be surprisingly well situated to consider new directions and new initiatives.

---

[6]See, for example, NAS, NAE, and IOM, *Science, Technology, and the Federal Government: National Goals for a New Era,* Washington, D.C.: National Academy Press, 1993; and NAS, NAE, and IOM, *Major Award Decisionmaking at the National Science Foundation,* Washington, D.C.: National Academy Press, 1994.

Legislative and executive branch response to the overarching issues identified in *Rising Above the Gathering Storm*[7] could foretell a brighter future. The President's American Competitiveness Initiative and the President's requested budget for FY2007 represent a welcome change in the funding outlook for elementary particle physics. The suggested increases in funding would help to enable the long-term vision for the U.S. program advocated in this report.

Multiyear Plans and Budgets

Many important experiments in particle physics require long-term investments and therefore multiyear plans and budgets. While the implementation of the priorities recommended below needs to be sensitive to budget realities, and also be sufficiently flexible to adapt to changes in the budget outlook, it is critical for the vitality of the U.S. program in particle physics to operate within the context of a long-range strategic plan. Indeed, in the FY2005 Energy and Water Development Appropriations Act, Congress directed DOE to develop a 5-year plan for DOE's Office of Science Programs, including the high-energy physics program. This plan enables program managers to develop more detailed and transparent multiyear plans.

The ability to make longer-term plans and commitments is also critical for international partnerships. One of the greatest challenges to U.S. leadership in future scientific activities, particularly in the case of particle physics, is to convince international colleagues that the U.S. political and budgeting processes are capable of sustaining the multiyear commitments that are negotiated when planning a joint venture. The sizeable U.S. investment in the LHC construction project at CERN (more than $500 million over 10 years) is an important demonstration that the U.S. particle physics program can make stable, long-term commitments.

Strategic Principle 5: Effective Long-Term Budget Planning. The Secretary of Energy and the Director of the National Science Foundation, working with the White House Office of Science and Technology Policy and the Office of Management and Budget and in consultation with the relevant authorization and appropriations committees of Congress, should, as a matter of strategic policy establish a 10- to 15-year budget plan for the elementary particle physics program.

---

[7]NAS, NAE, IOM, *Rising Above the Gathering Storm: Energizing and Employing America for a Brighter Economic Future*, Washington, D.C.: The National Academies Press, 2005 (Prepublication).

In the shorter run, given particular budgetary contingencies, the committee recommends that any necessary adjustment to plans be carried out in consultation with the research community.

NATIONAL PROGRAM CONSIDERATIONS

Strategic Principle 6: The Role of Fermilab. A strong and vital Fermilab is an essential element of U.S. leadership in elementary particle physics. Fermilab must play a major role in advancing the priorities identified in this report.

Particle physics benefits from close collaboration between universities and laboratories coast to coast. Over the years, each major laboratory involved in particle physics supported its own community of researchers both at the laboratory itself and at universities, creating a powerful synergy between these two communities that strengthened the national program. Fermilab has been no exception; research conducted at its Tevatron by laboratory staff and university collaborators has helped pave the way to the Terascale.

In recent years, however, the number of laboratories primarily devoted to particle physics has been shrinking. For instance, Brookhaven National Laboratory began to focus on nuclear physics in 1999. DOE's Basic Energy Sciences program became the major funder of SLAC in 2005. The current accelerator-based particle physics programs at Cornell, SLAC, and Fermilab are scheduled to be completed by 2009. After that, Fermilab will become the nation's only laboratory devoted primarily to particle physics. Continuing efforts at other major laboratories and from university groups will, however, be essential to realize the full potential of the nation's scientific agenda and regain the vitality and distinction of the U.S. program in particle physics. So, while a strong and vital Fermilab remains the essential element of U.S. leadership in this field, the overall program will also require a coordinated infrastructure of talent, resources, and leadership from other national laboratories and universities.

Whether or not it has an operating accelerator, Fermilab will be the focus of national efforts in Terascale physics, both by facilitating U.S. partnership in the LHC and spearheading U.S. participation in the ILC. It has the facilities, infrastructure, and intellectual capital needed to support U.S. particle physics, whether the experiments are conducted at home or abroad. There is no doubt that a distinguished national program requires a distinguished Fermilab. In addition, initial assessments of the area surrounding Fermilab indicate that it would satisfy some of the geological and environmental conditions required for the ILC, making Fermilab a natural choice for siting the ILC in the United States.

In any case, the committee expects that Fermilab will support and help mobilize the national program of particle physics research in the years ahead. In this new context, it is essential that Fermilab's internal priorities be aligned with those of the broader U.S. community.

Strategic Principle 7: The Advisory Structure in Particle Physics. A standing national program committee should be established to evaluate the merits of specific projects and to make recommendations to DOE and NSF about the national particle physics program in the context of international efforts.

The changing environment in particle physics requires a reexamination of the advisory structure for the field. In the past, individual national laboratories had their own program committees that provided advice to the laboratory directors on the feasibility of experiments and their prioritization within the laboratory's program. Ultimately, DOE in consultation with each director would approve the program for the laboratory, and DOE would provide the laboratory and associated university groups with the funding for the experiments. Overall program coordination has been facilitated by the HEPAP, a federal advisory committee originally charted in 1967; since 2000, it has been jointly chartered by NSF and DOE. As accelerator projects have become significantly larger and as more nonaccelerator programs have been proposed, a need has been recognized for a more comprehensive structure to establish national priorities in a time of tight fiscal constraints.

In November 2002, HEPAP implemented one of the central recommendations of its Long Range Planning Subpanel, established in 2001, to create the Particle Physics Project Prioritization Panel (P5). P5 was an ad hoc subpanel with a 2-year lifespan that has since been renewed. The tasks with which P5 has been charged have been changing over time as its responsibility grows. In early 2006, P5, together with its subpanels on dark energy, neutrino science, and other topics, was charged by the DOE Director of High Energy Physics and the NSF Assistant Director for Mathematical and Physical Sciences to develop and deliver to HEPAP a roadmap for particle physics.[8] While it is too early to tell whether this roadmap process will be successful, a higher level of analysis and overview of the U.S. portfolio is a step in the right direction.

The combination of unparalleled scientific opportunities and fiscal constraints will force the particle physics community to make some very hard choices. Under such circumstances, it would be enormously advantageous to have a national particle physics advisory apparatus that advises DOE and NSF on the scope of the U.S.

---

[8]HEPAP and its subcommittees are described on the DOE Office of High Energy Physics Web site at <http://www.er.doe.gov/hep/hepap.shtm>.

program and establishes priorities within the context of the international particle physics program. Its charge should be to evaluate the merits of specific proposals and make recommendations with regard to the national program to minimize unproductive duplication of overseas activities and to foster international collaboration for the benefit of science whenever practical. Such a coherent national advisory role could be played by an existing element of the high-energy physics program's advisory apparatus, such as a P5 committee that has been modified by transforming it into a standing committee with a broader mandate. The details of the specific advisory structure should be left to the agencies involved. Plans for rotating membership and participation from across the United States, as well as internationally, should be clear and public.

BUDGET CONSIDERATIONS

Within these strategic principles, different overall resource commitments can and must be accommodated. However, there is a point—a level of resources—below which a leadership is not tenable. Every effort should be made to avoid this situation, but if it nonetheless occurs, the strategic principles outlined above will need to be significantly amended. This is not, of course, an easy point to identify, but there is some initial evidence that the U.S. particle physics program is nearing it. Neither the nation's leaders in particle physics nor their sponsors have articulated or agreed on a compelling strategic plan that would sustain a distinctive leadership position for the U.S. program. That is, while the scientific community has identified the ILC as the highest priority project for the future, it has not succeeded in incorporating this element into a strategically focused program.

More generally, the current level of federal support for elementary particle physics presents an opportunity for strategic investment at the same time as it serves as an overall constraint. All priorities are set and the scope and timing of specific projects are decided with a budget in mind. The committee's principal recommendations assume that the current U.S. budget for elementary particle physics will at a minimum receive increases tied to the rate of inflation in the immediate future (Scenario A). This scenario would reflect a decision by policy makers to proceed with at least a constant level of effort,[9] although it implies that a smaller and smaller proportion of the U.S. gross national product would be

---

[9]The rate of inflation for scientific research and development, i.e., the growing cost of doing business in science, is a subject of much debate. Many have suggested that the scientific-research rate of inflationary growth is up to 2 percent higher than the usual Consumer Price Index metric. In the committee's analysis, Scenario A is properly defined as the constant-effort budget, thereby entraining the appropriate rate of inflation for scientific research.

devoted to this aspect of the scientific enterprise. The committee uses this particular scenario as the "control case" in recommending priorities for the next few years. However, even in this scenario it will be a significant challenge to sustain a position of leadership. Thus, the committee's initial set of recommendations and priorities assumes that there is a possibility of future growth in funding to allow for a critical major new initiative (Scenarios C and D). The possibility could be realized within the President's proposed FY2007 budget, which increases federal support for the physical sciences.

A scenario in which the existing budget remains flat without any adjustments for inflation was also considered (Scenario B). This scenario, which would reflect a decision by policy makers to continue to disinvest in this area of science, is incompatible with the desire to achieve a position of leadership for the U.S. program. It is the committee's view that such a policy would undermine any possibility that the United States will achieve a position of leadership as the committee has defined it. The consequences of such a decision for particle physics will be severe, and the implications for the nation's involvement at the frontiers of science and technology are equally sobering.

In Scenarios C and D, mentioned above, the current budget is increased annually in real terms (increases in addition to inflationary adjustments) by 2 to 3 percent (C) or by as much as 10 percent for the 7 years beginning in 2008 (D), as recommended in *Rising Above the Gathering Storm*. Both of these policies would reflect a national decision to increase the level of effort because the scientific opportunities in the physical sciences, especially in elementary particle physics, are currently so compelling. Chapter 3 detailed many of the discoveries that would be possible in Scenario C or Scenario D; many of these would not be possible in the constant-effort budget, Scenario A. The President's FY2007 budget request proposes an increase in funding for particle physics, which if implemented could be a step toward realizing Scenarios C or D. Figure 4-1 shows the four scenarios through the end of 2015.

The committee came to the alarming conclusion that for the United States to play a significant role in realizing the compelling science opportunities in elementary particle physics, the current short-term decline in inflation-adjusted resources devoted to this key area of science must be reversed as soon as possible. In the near term, funding levels should provide, at a minimum, a constant level of effort with perhaps some modest growth (Scenario A). Over the long term, a robust program will require leadership and real growth, somewhere between that posited for Scenarios C and D.

The committee is fully aware that real growth in the particle physics budget may take some time to be realized. It is essential, therefore, to reallocate those resources released from experiments scheduled to end in the next 3-4 years to fund

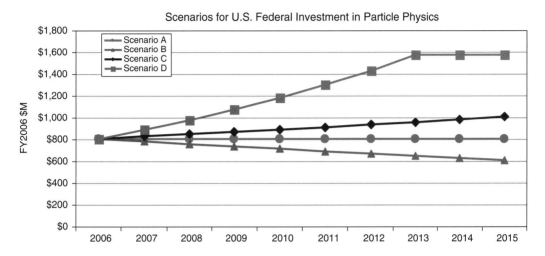

FIGURE 4-1 Comparison of out-year spending profiles in the four different budget scenarios considered by the committee. The profiles shown are in units of inflation-adjusted constant dollars assuming a future inflation rate of about 3 percent per annum. Although the committee's recommended strategy encompasses the next 15 years, this chart projects only the next 10 years because budget projections become quite unrealistic when looking a full 15 years into the future.

new elementary particle physics initiatives that address the most exciting scientific challenges. In a budget scenario that returns to real growth, it will be possible to achieve a position of distinctive leadership within a worldwide program and to support a diverse set of experiments that address current scientific questions more fully than is possible in the constant-effort scenario. As funding becomes available, either through new resources or from the conclusion of an existing activity, it should be allocated in accordance with the priorities outlined in this report.

In the long run, true leadership in particle physics will require augmentation of the resources devoted to the discipline. Simply put, the committee believes that the level of resources currently being committed to particle physics will be inadequate, in the long term, to obtain the technological, economic, social, and scientific benefits of undertaking the most compelling opportunities in this transformative area of science. Strengthening the U.S. role in particle physics will strengthen national and international confidence in the future of U.S. science and technology and in the image of the United States as a great nation supported by great science.

Finally, it is the committee's view that the competitiveness of the global and domestic economic environment the nation faces necessitates an aggressive investment in the mathematical and physical sciences, including particle physics, as well as in other areas of fundamental research.

Findings and Recommended Actions

THE SCIENTIFIC AGENDA FOR ELEMENTARY PARTICLE PHYSICS

New technological capacities now make it possible to address long-stand ing questions in particle physics that are encompassed within the follow ing three questions:

- Can all the forces between particles be understood in a unified framework?
- What do the properties of particles reveal about the nature and origin of matter and the properties of space and time?
- What are dark matter and dark energy, and how has quantum mechanics influenced the structure of the universe?

Two special considerations give these questions an urgency. First, a rare opportunity currently exists for the U.S. program to collaborate with international partners to transform today's understanding of all three of these questions. This window of opportunity will not remain open for long. Second, the U.S. effort in particle physics has been without a compelling, clearly articulated, and widely held strategic vision since the cancellation of the Superconducting Super Collider, and the lack of such a vision has now become critical. The committee's recommended agenda for the U.S. role in particle physics addresses both considerations.

The committee's recommended agenda not only seizes the tremendous opportunity for intellectual transformation as the discoveries of the Terascale become available, but it also calls for a transformation in how particle physicists interact with one another at the national and international levels; this turning point in particle physics is extremely compelling.

PRIORITIES

In considering the actions recommended in this chapter, it is important to understand what the committee means by "priority." The elements of the scientific agenda that it recommends have been prioritized based on its analysis of the importance of the underlying scientific opportunities combined with its assessment of the technical readiness and feasibility of experimental facilities located in the United States and abroad.

It also is important to keep in mind the strategic principles outlined in Chapter 4. In particular, it is important to recall the strategic necessity of mounting, regardless of budgetary constraints, a comprehensive program that reflects a diversity of scientific opportunities and approaches to the scientific challenges facing particle physics. Under no circumstances, therefore, should the committee's top two or three priorities be permitted to exhaust the entire available budget. Indeed, in the most pessimistic budget scenario, where maintaining a position of leadership is unrealistic (Scenario B), the resources invested in the priorities outlined below would need to be adjusted, but the need for pursuing a diversified research portfolio would be unchanged.

The capacity of the Large Hadron Collider (LHC) and the International Linear Collider (ILC) to explore the Terascale directly offers the promise of deep insights into such matters as the Higgs boson, supersymmetry, dark matter candidates, and hidden spatial and quantum dimensions. At the same time, explorations of unification and particle astrophysics, including both space-based and underground observations, promise to shed light on dark energy, dark matter, and inflationary models of the universe. Moreover, new and planned precision studies of lepton and quark properties and their interactions may reveal the role of neutrinos in the universe, explain how matter came to dominate antimatter, or uncover entirely new phenomena. The committee discusses each of these scientific opportunities and the associated action items in priority order, assuming, for the moment, the constant-effort budget (Scenario A).

DIRECT EXPLORATION OF THE TERASCALE

The most compelling current scientific opportunity in elementary particle physics is aggressive exploration of the Terascale, and this is the committee's highest priority for the U.S. program. A two-part strategy using two world-leading accelerators is required, as described below. Direct investigation of phenomena at the energy frontier holds the greatest promise for transformational advances. Realizing the scientific potential of the Terascale requires experiments using particle accelerators. Within this context, the LHC and the proposed ILC experimental programs offer the best routes. The committee's recommended strategy is also predicated on the observation that, at the current time, a higher risk, higher reward strategy is necessary if the United States is to sustain a leadership position in the decades ahead. To accomplish this, our nation must take the initiative in aggressively exploring the most compelling new science opportunities.

Finding 1: The LHC Experimental Program. The study of LHC physics will be at the center of the U.S. particle physics program during the coming decade.

The LHC is scheduled for start-up in 2007 at CERN and will be at the center of exciting developments in elementary particle physics over at least the next 15 years. By colliding particles at the TeV energy scale (the Terascale), the LHC will provide the first look at a landscape expected to be rich in answers to questions about the origins of mass, hidden dimensions, and the limits to current understandings of the quantum universe. As U.S.-based facilities conclude their particle physics programs over the next few years, more and more U.S. scientists and students, as well as many others from around the world, will be focusing their efforts at the LHC, which soon will become the center of gravity for experimental particle physics.

The United States has already made a very substantial contribution of human and financial capital to the development of the ATLAS and CMS detectors at CERN, as well as to construction of the accelerator, but it is critical to adequately support the U.S. research groups that will carry out experiments at the LHC and to participate in the upgrades of the LHC's experimental facilities. In addition, U.S. centers for data analysis have been set up at Fermilab and Brookhaven National Laboratory, as have smaller data centers at other national laboratories and universities, and these critical analysis and computing facilities must be supported during the LHC's lifetime.

The committee notes the importance of devoting resources to the continued development of cyberinfrastructure if the United States is to play a leadership role in the LHC program. In 2007, for the first time, the United States will enter an era when its primary experimental research in particle physics will be based in a foreign country. Careful attention to networking, data access, and collaborative tools will be required to ensure the full participation of U.S. scientists and students in the global LHC effort and realization of the scientific opportunities offered by the LHC.

As the LHC physics program unfolds over the next decade, proposals for upgrades will likely become better formulated. For instance, the detectors might be upgraded to cope with increases in accelerator event rates, or the accelerator might be significantly upgraded to double the energy. Early discoveries at the LHC are likely to guide the pace of the surrounding discussions.

Action Item 1: The LHC Experimental Program. The highest priority for the U.S. national effort in elementary particle physics should be to continue as an active partner in realizing the physics potential of the LHC experimental program.

The number of U.S. researchers working at the LHC will continue to increase as operations begin and the earliest results of Terascale science appear. As the most immediate scientific opportunity for U.S. researchers, the LHC should receive U.S. support in line with the level of involvement of U.S. researchers and the need to maintain the detectors. The committee expects that full support of the growing U.S. participation will consume a larger share of resources than at present. As potential upgrades to the detectors and the accelerator are motivated and defined by scientific results, the U.S. particle physics program should consider in-kind contributions where appropriate. With regard to LHC accelerator operations, direct funding should only be considered within the context of international discussions that address the broad suite of international scientific collaborations both inside and outside particle physics.

Finding 2: Achieving Readiness for the ILC. An aggressive approach to realization of the ILC needs to be the central element in a new strategic plan for the U.S. program in particle physics.

Exploration of the Terascale is the highest priority for particle physics, and a linear collider is the next critical element required to meet this objective and carry the science program well beyond the next decade. The science program for the proposed ILC addresses the major contemporary challenges in particle physics and extends the discovery reach for Terascale physics. The ILC is therefore the most important new experimental facility in elementary particle physics. It is envisioned to have a total energy of 500 GeV during its initial phase, with a planned capability for a subsequent increase to 1 TeV. It is to discover the nature and meaning of results from the LHC that an electron-positron collider such as the ILC is required. As a result, the elementary particle physics community worldwide is in consensus that the ILC should be the next major experimental facility to be built for particle physics. Furthermore, the scientific, technological, and industrial expertise needed to build and operate the facility is becoming available, and complete capabilities will fall within reach alongside a comprehensive R&D effort.

The study of physics at the ILC will be at the center of the U.S. particle physics program beyond 2015.

Additional R&D is necessary to resolve the remaining technological issues and to formulate a set of design and manufacturing requirements that will minimize the cost of this multi-billion-dollar facility. The Global Design Effort (GDE) currently under way expects to prepare the Reference Design Report (RDR) with a baseline cost estimate by the end of 2006, with a full technical design ready in 2009. The GDE is currently setting the strategies and priorities for the work of hundreds of scientists and engineers at universities and laboratories around the world. More than $50 million was expended for these efforts last year in Europe, a similar amount was spent in Japan, and $25 million in the United States. This research is essential to reduce the technical and cost risks before such a project can be approved.

Clearly, a project as large and complex as the proposed ILC can be pursued only if an international consortium can be formed to pursue the design and remaining R&D and to devise a fair mechanism for sharing the costs, scientific leadership, and participation in the ILC program. While there is a great deal to be learned from existing models of international collaboration—examples include ITER, the Atacama Large Millimeter Array (ALMA), and CERN—the ILC may require a unique form of collaboration. To date, international collaboration on the scientific and technical issues surrounding the ILC has been excellent, as shown by the consensus on the ILC technology selection, ongoing R&D collaboration, and initiation of the GDE. (Appendix A describes some of this progress.)

Strong theoretical arguments and past experimental results provide convincing evidence that the Terascale will offer a rich spectrum of physics that demands exploration by both proton and electron colliders.[1] An informed decision on the construction of the ILC could be made as soon as a credible cost estimate exists and an appropriate governance structure takes shape; ideally it should be made no later than 2010, by which time the LHC should have revealed some of the nature of the new physics that lies at the Terascale. Timely and responsive decisions on the ILC will optimize the forward momentum and continuity of the field's research pursuits. This time frame is compatible with the expected start-up date of the LHC and with the conclusion of current accelerator-based experimental programs in the United States.

---

[1]See, for example, LHC/ILC Study Group, *Physics Interplay of the LHC and ILC*, 2004, available online at <http://arxiv.org/abs/hep-ph/0410364> (last accessed February 1, 2006).

Action Item 2: Achieving Readiness for the ILC. The United States should launch a major program of R&D, design, industrialization, and management and financing studies of the ILC accelerator and detectors.

U.S. expenditures on R&D for the ILC should be very significantly expanded. The key objective of this R&D program is to reduce both the technical and cost risks of the ILC and to initiate a program that will allow for industrialization of significant portions of ILC components. This effort should continue in the tradition of the broad international collaboration that has been the hallmark of the ILC project to date. The United States should prepare for long-term involvement in the physics program of the ILC as well.

For the accelerator, this commitment should be at a level as high as $100 million in the peak year and could represent a cumulative amount on the order of $300 million to $500 million over the time period prior to the decision to proceed with construction.[2] For detector R&D, the commitment would be near $80 million over the same period,[3] financed in part by the redirection of some university and national laboratory efforts.

The committee believes strongly that it is in the best long-term interests of the U.S. program in particle physics to make a significant investment in this R&D program and to become a leading center for ILC R&D well before a construction decision is made. This is a critical element of the committee's recommended strategy: The United States must take the initiative now to be in a position to make a

---

[2]These levels of effort are based on tables provided to the committee by the GDE and the U.S. Linear Collider Steering Group and the High Energy Physics Advisory Panel on May 15, 2005, in response to a written request from the committee for information. With certain adjustments and small differences in emphasis, this investment profile is in agreement with expectations from the GDE for the necessary worldwide R&D to prepare for a construction decision. The committee reviewed the proposed investment profile carefully and used its best judgment to characterize the entire planned effort with robust figures. Finally, the proposed schedule does separate out the "globally shareable investment" in ILC research and development and the additional investment required for the United States to provide and certify a site for a bid to host the ILC (for instance, the analysis includes figures for meeting environment, safety, and health labor regulations; budget management; and contingency). Accepting the current cooperation brokered by the GDE, the committee posited that the United States would partake in an equal third of the globally shareable investment in addition to the expected costs for developing a U.S. site. Based on its collective experience in project management and cost projections, the committee came to agreement on the proposed range and schedule of investment as proposed in the text.

[3]These numbers are based on an updated report by the ILC Detector R&D Panel: J.-C. Brient et al., *ILC Detector Research and Development: Status Report and Urgent Requirements for Funding*, January 2006, available online at <http://physics.uoregon.edu/~lc/wwstudy/R&D%20Report-final.pdf> (last accessed July 1, 2006).

credible bid to host the ILC when the R&D effort is complete. Clearly, however, the decision to proceed with actual construction would require the establishment of an international governance structure, an international decision on a site for the ILC, and reliable and robust cost estimates.

The ILC project has been conceived and planned in a manner very different from the Superconducting Supercollider (SSC) project of the late 1980s and early 1990s. The committee concurs that successful implementation of the ILC will require different approaches than were employed with the SSC. The ILC planning and R&D activities to date have been managed by the scientific community as a truly international effort. These activities have laid a solid foundation for the eventual development of an international governance structure and cost-sharing agreement, both of which need to be in place prior to the start of construction. The committee also concurs with the need for a rigorous R&D and industrialization program before a decision to construct in order to minimize technical risks and uncertainties in cost and schedule.

In addition, the committee believes that the U.S. effort to become the host site for the project should take full advantage of the existing infrastructure and expertise of Fermilab. Using Fermilab as the host laboratory would avoid start-up issues associated with a new site like those that affected the SSC. The committee also notes that DOE has adopted a new project management system since the SSC (under DOE Order 413) that strengthens oversight and cost management of major projects. In fact, since the mid-1990s, the Office of Science at DOE has a remarkably good record of meeting performance, schedule, and budget targets.[4] The committee expects that the United States and its international partners will employ state-of-the-art information systems to aid in the management and reporting of project implementation activities. However, the committee anticipates that some procedural aspects of the DOE project management system may need to be modified to accommodate international participation in the project. This issue should be addressed in more detail as part of the planning for the United States to bid to host the ILC.

Finding 3: The Benefits of Hosting the ILC. Hosting the ILC would inspire students, attract talented scientists from throughout the world, create a suite of high-technology jobs, and strengthen national leadership in science and technology.

---

[4]For historical reference, see L. Edward Temple, Jr., "Office of Energy Research Project Performance," Department of Energy, Office of Energy Research, Construction, Environment, and Safety Division, February 1986. Project performance since 2000 has been guided by DOE Order 413.3, and some evaluations are available at the DOE Office of Science's Office of Project Assessment, available at <http://www.science.doe.gov/opa/>.

The ILC will be a flagship scientific facility. It will focus on some of the most profound and mind-stretching questions in science. This challenge will fire the creativity and imagination of many of the nation's brightest young minds. Hosting this exciting project can be expected to increase student interest in science and engineering and thus enhance the nation's scientific and technological workforces, just as many of today's engineers and scientists were attracted to these fields by the nation's commitment to the space program in the 1960s and 1970s.

In addition, the ILC will attract thousands of talented scientists and students from around the world. If history is any guide, many of these highly talented and motivated scientists will remain in the United States and continue to contribute to the nation's technological leadership, which in turn will stimulate domestic economic growth through scientific and technological innovation. Moreover, some of the world's best scientists undoubtedly will join the nation's universities to be close to the project, thereby enriching these institutions. A key component of the success of U.S. academic research institutions has been direct and easy access to the world's premier research facilities and infrastructure; hosting the ILC would extend this pattern of success into the 21st century. In short, constructing and operating a world-class facility will create an unparalleled intellectual environment and stimulate innovation and creativity.

The nation to host the ILC will require a substantial number of highly trained accelerator physicists, engineers, and technicians to operate the facility. The ILC will be one of the world's premier training facilities for bright young people entering accelerator and particle physics. Many of these young accelerator physicists will take their skills to other areas of science and technology such as biomedical applications and materials studies and fabrication. This could contribute to the creation of jobs across the nation, not only in high-technology sectors but in all sectors that benefit from a strong economy and the creation of knowledge.

Finally, investments by other nations in the facility will be very significant, principally through in-kind contributions. These contributions can add value to the host nation by leveraging the skills and abilities of the U.S. technical workforce and industry.

The intellectual benefits of hosting a flagship scientific facility are illustrated by the LHC at CERN. Although the United States is an important partner in this enterprise, the scientific soul is in Geneva. To participate in and significantly contribute to the scientific program of the LHC, U.S. researchers regularly travel (and will continue to travel) to Switzerland to conduct their research on-site at the LHC. Furthermore, scientists directing graduate students find it increasingly important for their young particle physicists to spend a year or more at an operating experiment to fully appreciate and understand the principles that connect day-to-day operations with the underlying quest for physics.

If the United States is successful in attracting the ILC, the actual construction and operation of the facility will require not only international partners but an increase in the resources devoted to the U.S. program. The constant-effort budget underlying the control scenario (Scenario A) will not be sufficient to fully fund the U.S. share of the construction and operation of the ILC's accelerator and detectors.

Although the U.S. community of particle physicists, accelerator scientists, and engineers does not now have the capacity to undertake an effort as large as the construction and operation of the ILC, there is every expectation that the excitement of such an initiative would attract more than sufficient talent to the field to make it possible by the time a construction decision is announced. That is, the U.S. physical science and engineering community has a lot of latent expertise that could be tapped for the ILC and, the committee believes, would allow the United States to host the ILC.

It is not certain that the aggressive pursuit of both ILC R&D and a bid to host the facility would ensure that the project moves toward construction or that the site chosen would be in the United States. It is the committee's view, however, that deferring or avoiding an investment in R&D and mounting a compelling bid to host the ILC will not achieve a leadership position for the U.S. program, even if it might be less risky. The committee explicitly acknowledges that focusing U.S. efforts on such a major and singular enterprise would expose the U.S. particle physics program to risk, and while various creative strategies might mitigate the risk, it cannot be entirely avoided.

The decision to take an aggressive approach to the realization of the ILC does not in and of itself constitute a compelling strategy for the U.S. particle physics program. However, it is an essential element in any strategy based on an initial constant-effort (or better) budget and would provide an opportunity for the U.S. program to achieve a well-defined and distinct position of leadership in the next 15 years. The committee strongly believes that the risk-adjusted return on an investment that enables the United States to become a major center for ILC R&D and to prepare a bid to host it is the best chance the U.S. particle physics program has to occupy a distinct position of leadership over the next decade and beyond.

Action Item 3: The Path Forward for the ILC. The United States should announce its strong intent to become the host country for the ILC and should undertake the necessary work to provide a viable site and mount a compelling bid.

As the host country for the ILC, the United States would likely need to commit to a higher cost share than its international partners. Based on experience, it is expected that 30 to 35 percent of the total cost would be for conventional construction activities at the site. As part of the planning for hosting the project, the DOE and NSF should undertake a study of alternative methods for financing the cost of conventional construction work at the host site. These alternatives could include greater assurance for public appropriations (e.g., lump-sum or advance appropriations), nonfederal contributions, and federally backed third-party financing. Such methods would smooth the construction budget profile and could involve financing program expenditures over multiple years. (For instance, in accord with Strategic Principle 5, a construction project might arrange third-party financing by future planned appropriations.) Experience with science projects shows that uncertainties and shortfalls in annual appropriations can be a leading cause of unnecessary cost escalations and inefficient and unwise, though expeditious, decisions. Alternative methods of financing conventional construction could provide greater funding stability and, in turn, greater certainty that cost and schedule goals will be achieved. Reliance on annual federal appropriations to finance these costs could require a significant increase in the annual budget for the U.S. particle physics program when construction begins.

Locating the ILC near Fermilab would give U.S. elementary particle physics a vibrant center in the coming decades. The existing infrastructure at Fermilab, as well as initial assessments of its geological stability and useable space, make it a logical choice for the ILC in the United States. As the only national laboratory devoted solely to particle physics, Fermilab's top priority should be to secure the ILC.

One issue that the committee did not address in its analysis was a detailed cost estimate for the ILC.[5] The committee was aware that several preliminary estimates had been developed in the United States and elsewhere, but it concluded that these estimates were based on different design concepts and did not necessarily reflect the current plan for the project. The committee also has monitored closely the ongoing GDE, which is currently scheduled to produce an RDR by the end of 2006 that will include an RDR-based preliminary cost estimate. Successful completion of the RDR exercise will be an important demonstration of the feasibility of an ILC. The committee recognizes the prudence of this approach: A credible estimate of project cost must await a specific set of design parameters and, later, the international selection of a viable site. In general, the committee notes that the scale,

---

[5]The committee reminds the reader that the formulation and development of a large-facility project is defined by two elements: a technical definition or scope of work and an implementation plan. A project cost is not logical or credible until these two have been developed and agreed on; as the scope and schedule change, the cost will also evolve.

complexity, and engineering challenges of the ILC are expected to be very roughly comparable to those associated with the LHC.[6]

The committee's recommendations on the path forward toward the ILC are based on the premise that the GDE will produce an RDR that is acceptable to decision makers in the United States and other countries as a basis on which to proceed. This statement may appear to indicate only conditional support for the ILC, but the opposite is true: The committee believes that the ILC is such a tremendous opportunity that it must be pursued vigorously and wholeheartedly. The committee's recommendations are intended to help guide the next phases of the R&D and the detailed design and international collaboration processes. They are also intended to form a set of expectations in advance of a decision to proceed with the project.

The elements of the committee's highest priority recommendations and the relationship among them may require additional explanation. Exploration of the Terascale is the committee's highest priority because it offers the most compelling science opportunities. This exploration will begin at the LHC within 2 years, while the detailed explorations of the Terascale at the ILC are perhaps a decade or more away. Exploiting the LHC and taking a leadership role in ILC R&D must proceed along different paths of action because the construction phase of the LHC is essentially complete and the global particle physics community is ready to conduct experiments with it, whereas the ILC is in an embryonic stage. The United States has invested heavily in the LHC, and U.S. scientists are preparing for operations. The effort to design and build the ILC has, relatively speaking, just begun, with the GDE representing the first phase of an internationally (but informally) coordinated program; the creation of an institution to oversee the building of the ILC is still in the future. The committee strongly believes that a firm commitment to the R&D phase of the ILC and the development of a bid to host the project are necessary to give the United States a scientifically advantageous position when the construction decision becomes tenable.

In summary, within the strategic principles outlined in Chapter 4, the committee believes that the two highest priorities for the U.S. particle physics program are aggressive support for the LHC and ILC programs. In the peak years of the planning and R&D for the ILC accelerator and detectors, support for the LHC and ILC experimental programs will demand a large fraction of the U.S. particle physics budget.

---

[6]By DOE estimates, "the total cost of the LHC [accelerator] on a basis comparable to that used for U.S. projects is estimated at about $6 billion," excluding the cost of excavating the underground tunnel. See DOE FY2006 Congressional Budget Request, p. 287, for more details.

EXPLORATIONS OF PARTICLE ASTROPHYSICS AND UNIFICATION

Finding 4: Opportunities at the Interface of Particle Physics, Astrophysics, and Cosmology. Elementary particle physicists have an extraordinary opportunity to make breakthrough discoveries by engaging in astrophysics and cosmology research that probes energies and physical conditions that are not available in an accelerator laboratory. The investigations simultaneously search for new laws of nature and advance understanding of the origin, evolution, and future of the universe.

The United States has already established a leadership position at the interface of particle physics, astrophysics, and cosmology, but with international activity growing rapidly, further investment will be needed to maintain that leadership.

The committee has identified three major research challenges that are ripe for pursuit:

- The direct detection of dark matter in terrestrial laboratories, the results of which could then be combined with measurements of candidate dark matter particles produced in accelerators.
- The precision measurement of the cosmic microwave background (CMB) polarization, which would probe the physics during the inflation that appears to have occurred within a tiny fraction of a second following the big bang.
- The measurement of key properties of dark energy.

Action Item 4: Coordination of Efforts at the Interface of Particle Physics, Astrophysics, and Cosmology. Scientific priorities at the interface of particle physics, astrophysics, and cosmology should be determined through a mechanism jointly involving NSF, DOE, and NASA, with emphasis on DOE and NSF participation in projects where the intellectual and technological capabilities of particle physicists can make unique contributions. The committee recommends that a larger share of the current U.S. elementary particle physics research budget should be allocated to the three research challenges articulated above.

NASA has historically played a critical role in this area, and it should continue to do so. Projects that cut across agencies and research communities require an additional level of planning and coordination to ensure success, especially when multiple research communities are involved that have overlapping scientific priorities. The key element is coordination of efforts to respond to scientific opportunities. For instance, dark energy can be explored from space and from the ground, and researchers from particle physics, astrophysics, and cosmology have all expressed interest in doing so. Ways to explore dark energy should be pursued jointly by all three constituencies and the three agencies.

There are existing mechanisms that could provide such coordination. A good example is the broader astronomy and astrophysics decadal survey process, which has provided strategic advice to NSF and NASA in the form of a list of scientific priorities for each decade.[7] Not all opportunities at this interface require the simultaneous involvement of all three agencies, of course. Typically, ground-based projects, such as the Large Synoptic Survey Telescope, require coordination between DOE and NSF, while space-based projects, such as the Joint Dark Energy Mission, require coordination between DOE and NASA. Between DOE and NSF, both HEPAP and P5 have started taking an active role in providing coordination of the joint portfolio at this scientific interface. Finally, the Astronomy and Astrophysics Advisory Committee (AAAC), chartered by Congress in the NSF Authorization Act of 2002, is a relatively new mechanism that has started to provide tactical guidance to all three agencies about the implementation of joint projects.

Since current commitments from the particle physics budgets to opportunities at the interface are relatively modest compared to the full particle physics program, it is the sense of the committee that they should be built up to two to three times their current level.

Finding 5: Probes of Neutrinos and Proton Decay. A program of neutrino physics including, eventually, a detector sensitive enough for proton decay offers a probe of unification physics.

In the past 10 years, it became clear that neutrinos have tiny but nonzero masses. This is a departure from the Standard Model and may be a signal of the unification of particle forces.

There are now opportunities to extend this hint of unification. Proton decay experiments might show that the proton is unstable, a monumental discovery that would confirm one of the most basic predictions of unified theories. Neutrinoless double-beta decay experiments could demonstrate that the neutrino is its own antiparticle, which would greatly strengthen the case for interpreting neutrino masses in terms of unification. Experiments that measure the neutrino mixing angle θ_{13} and the CP-violating parameter that affects neutrino oscillations could provide additional information about particle unification. Finally, important clues

[7]NRC, *Astronomy and Astrophysics in the New Millennium*, Washington, D.C.: National Academy Press, 2001.

about unification could come from other experiments, including observation of the polarization of the CMB (one of the particle astrophysics priorities recommended above) and axion searches.

A recent study of neutrino physics by the American Physical Society[8] identified a set of important questions that need to be addressed and laid out a progressive program of research. The two highest priority recommendations were (1) to establish whether or not the neutrino is its own antiparticle through a phased program of neutrinoless double-beta decay experiments and (2) to carry out a program of experiments to establish the remaining parameters in the neutrino mixing matrix with the goal of understanding CP violation for neutrinos. The second program depends on the value of θ_{13}. It could include reactor experiments sensitive to θ_{13}, long-baseline accelerator experiments sensitive to θ_{13} and capable of determining the ordering of neutrino masses by observing matter effects, and, eventually, a large-scale, long-baseline experiment with a large multipurpose underground detector capable of detecting CP violation. This large underground detector also could search for proton decay. DOE recently announced a set of mission needs in neutrino physics along the lines recommended by the APS study.

Full exploitation of neutrino physics requires diverse modalities of experimentation, including accelerator beams, reactors, underground experiments, and, especially, underground neutrinoless double-beta decay experiments. Significant efforts are under way in Asia and Europe, as well as in the United States and Canada. In the United States, some neutrino experiments are supported by the nuclear physics community, so effective coordination is essential within DOE's Office of Science. The NSF is overseeing a process to develop proposals for a U.S. deep underground science and engineering laboratory (DUSEL) that would provide scientists from physics, geology, and biology with a technical infrastructure to conduct investigations deep underground. Neutrinoless double-beta decay and proton decay experiments are examples of projects that could take advantage of such a facility.

Action Item 5: A Staged Neutrino and Proton Decay Research Program. The committee recommends that the properties of neutrinos be determined through a well-coordinated, staged program of experiments developed with international planning and cooperation.

• **A phased program of searches for the nature of neutrino mass (using neutrinoless double-beta decay) should be pursued with high priority.**

---

[8]American Physical Society, *The Neutrino Matrix*, Washington, D.C., 2004.

- DOE and NSF should invite international partners in order to initiate a multiparty study to explore the feasibility of joint rather than parallel efforts in accelerator-based neutrino experiments. Major investments in this area should be evaluated in light of the outcome of this study.
- Longer-term goals should include experiments to unravel possible CP violation in the physics of neutrinos and renewed searches for proton decay. There may be a valuable synergy between these important objectives, as the neutrino CP violation measurements might require a very large detector that, if placed deep underground, would also be the right instrument for detecting proton decay.

The committee believes that in order to give U.S. researchers access to the best scientific opportunities in a timely manner, international cooperation and coordination are essential from start to finish. Existing individual projects in these areas, especially neutrinoless double-beta decay, exemplify strong and diverse international participation. The committee recommends, however, that experiments in this area should be globally rationalized from start to finish. Facilitating such a rationalization process is a necessary part of the U.S. commitment to leadership. The effort would also work to avoid unnecessary duplication and would most efficiently deploy the worldwide investments in the field. For instance, efforts are under way in Japan to finish the construction of a high-intensity proton source that has important applications for certain neutrino experiments; the United States is exploring opportunities for experiments that could use different baselines to achieve different sensitivities; and there are proposals for neutrino beams at CERN. The objective of investigating the feasibility of a joint program is not simply to avoid unnecessary overlap or duplication of experiments. Rather, in the constrained budget environment facing the international particle physics community, it is to explore whether pooling resources can lead to a more robust scientific program and achieve key experimental results more quickly.

Finding 6: Precision Probes of Physics Beyond the Standard Model. Studies of the patterns of weak interactions (particularly rare decays and CP violation in the quark sector), dipole moments, tabletop tests of gravity, and lepton flavor and lepton number violation could expand our understanding of and more precisely define the physics that might lie beyond the Standard Model.

The information from such studies is complementary to that obtainable from direct searches for new particles at the LHC and ILC and has historically played an important role in constraining models of new physics. The current B factories and CLEO-c will conclude their programs in this arena by the end of 2008. Future B physics efforts include the LHCb and a possible future super-B factory (under consideration in Japan and Italy). Lepton flavor violation studies offer an important window on new physics, as do searches for rare kaon decays. Some of these studies require meson beams from a proton facility, such as J-PARC in Japan; others can proceed using a high-intensity electron-positron collider, such as the Beijing Electron-Positron Collider facility in China. Precision measurements of the muon g-2 parameter and searches for electric dipole moments also offer new constraints on physics beyond the Standard Model. Some of the latter can be relatively small-scale efforts; experiments with significantly improved reach are possible within the next few years.

> **Action Item 6: Precision Probes of Physics Beyond the Standard Model. U.S. participation in large-scale, high-precision experiments that probe particle physics beyond the Standard Model should continue, but the level of support that can be sustained will have to be very sensitive to the overall budget picture. Only very limited participation will be feasible in budget scenarios with little or no real growth. Participation in inexpensive, small-scale, high-precision measurements should be encouraged in any budget scenario.**

This is an area where investment and collaboration in joint international projects can offer significant opportunities and leverage to all parties. Small-scale experiments should be supported as part of the overall program when they offer significant reach into unexplored physics.

IMPLICATIONS OF THE STRATEGIC AGENDA UNDER
DIFFERENT BUDGET SCENARIOS

The committee recognizes that the United States could pursue more than one strategy in particle physics in the next decade. It outlines here the strategy that will have the highest risk-adjusted return and, in its best judgment, will be most likely to sustain U.S. leadership in particle physics. Failure to participate in active exploration of the Terascale would surrender U.S. leadership because the nation would not be able to pursue the most compelling science opportunities. Finally, the committee considered strategies that abandoned accelerators altogether; these approaches were rejected because they did not lead to a national program of sufficient health and vitality to sustain itself.

As the committee considered the constant-effort budget (Scenario A), it estimated that such a budget could support the LHC and ILC efforts it recommends, as well as an expansion of the current efforts in particle astrophysics through 2010. Moreover, as long as the program adheres to the strategic principles outlined in Chapter 4, there would be adequate funds to support some smaller programs that are important to the field. However, the Scenario A budget would not provide much funding for neutrino-mass measurements, nor would it allow proceeding with any major accelerator-based neutrino program based in the United States without significant foreign contributions. The more ambitious initiatives in neutrino physics would need to rely on forging alliances with colleagues abroad.

The straightforward set of priorities articulated in action items 1 through 6 could take the U.S. program through the next 5 years, but if a decision is made to go ahead with the proposed ILC, important new considerations will enter the picture. Most important, the ILC is a multi-billion-dollar facility, and if the United States were to be the host, as the committee has recommended, it would be expected to shoulder a significant fraction of the costs. While the ongoing U.S. program could and should provide a significant share of the necessary funds, it cannot fully cover the expected contribution of the host country. Funding beyond that assumed in Scenario A would be required to build and operate the ILC.

The committee also considered the appropriate set of priorities under Scenario B. In this more pessimistic case, the committee still recommends that the highest priority be participation in the LHC and the ILC programs over the next 5 years, so that the U.S. particle physics program could provide U.S.-based scientists and students with opportunities to participate in the most exciting aspects of elementary particle physics. However, the committee believes that in Scenario B the United States would not be able to host the ILC; rather, it would be a strong participant in such a facility hosted abroad. Indeed, in Scenario B, full U.S. participation in the exploration of the Terascale might well be jeopardized. In this scenario, the United States could expand its efforts in particle astrophysics and participate in globally coordinated neutrino experiments abroad. However, distinctive and distinguished U.S. leadership in particle physics would very likely be sacrificed. Certainly the scientific influence of the U.S. program in particle physics would be much diminished. Even with an ILC located overseas, U.S. participation would require budget increases well above the Scenario B level after the initial 5 years.

It was apparent to the committee that Scenario C could provide a significant portion of both the capital and operating resources necessary for the United States to host the ILC. This goal and participation in the LHC would remain the top priorities. However, this scenario would allow an expanded program in particle astrophysics (the committee's next highest priority) and a fuller program with international collaborators in neutrino physics and proton decay and in flavor physics. This scenario would go a long way toward ensuring retention of the infrastructure and expertise required for the ILC and toward securing a U.S. presence among the leaders in this field. The more optimistic scenario, D, would offer opportunities to engage strongly in all aspects of the science described in this report and to recover more securely the U.S. role as a leader in this field. Under this scenario, operation of a facility on the scale of ILC would probably not require additional funding, though additional support during construction might still be needed.

REALIZING THE STRATEGIC VISION FOR ELEMENTARY PARTICLE PHYSICS

As some important elementary particle physics experiments at U.S. national laboratories complete their current objectives, Fermilab will have a special role as the only national laboratory devoted completely to particle physics. In framing the future role for the United States in elementary particle physics, the committee would like to emphasize not only the importance of a flourishing and dynamic Fermilab but also the necessity of ensuring that the overall resources of the U.S. particle physics community be deployed to best effect.

Implementing the committee's recommended priorities will require strong leadership and a strong commitment to a common vision from all stakeholders (including the particle physics community, federal research agencies, the Office of Science and Technology Policy, the Office of Management and Budget, and Congress). Previous planning efforts were sometimes not fully realized because they lacked coherence, a clear consensus on relative priorities, and a commitment to implementation. Moreover, in the past, planning at the national laboratories was not tied to an overall national plan. When adequate resources were available, this democracy of ideas was a strong point of the program; now, somewhat more centralized planning and implementation are necessary. To move forward, the community must adopt a new way of making decisions, and those decisions need to follow from a strategic plan.

Elementary particle physics is poised to make potentially transformative discoveries. If the United States commits to a strategic vision such as the one the committee has laid out, the nation can continue to occupy a position of leadership in this vibrant and exciting science. Such an aspiration is worthy of a great nation wishing to remain on the scientific and technological frontiers. It will inspire future generations, repay the investments many times over, and provide a fuller understanding of mankind's place in the cosmos.

Afterword

Every 10 years or so, the National Research Council's Board on Physics and Astronomy (BPA) engages in a decadal survey of physics. The current survey, *Physics 2010*, is under way and is expected to be completed over a 5-year period following its inception in 2005.

The *Physics 2010* decadal survey is focused on an assessment of and outlook for each branch of physics. Each assessment will be conducted by an independent study committee appointed by the National Research Council based on the advice and recommendations of the BPA. This decadal survey of physics serves two broad purposes: (1) it provides a periodic snapshot of the field that is useful for tracking and understanding the evolution of the science, and (2) it provides a process whereby compelling emerging opportunities can be identified and developed.

The *Physics 2010* project will include reports on atomic, molecular, and optical science; plasma physics; condensed matter and materials physics; elementary particle physics; and nuclear physics. The Committee on Elementary Particle Physics in the 21st Century undertook the preparation of this volume, the first of this series.

Appendixes

A

International Progress
Toward the ILC

Building on separate regional efforts, several studies to develop a path for ward for the ILC were initiated under the auspices of the International Committee for Future Accelerators (ICFA), with the initial technical study issued in 1995. Since then, the ILC concept has successfully passed through a number of key milestones, including these:

- In August 1999, ICFA issued a statement concluding that a linear collider would produce compelling and unique scientific opportunities, and it recommended vigorous pursuit of R&D on a linear collider.
- In 2001, the U.S. DOE/NSF High Energy Physics Advisory Panel (HEPAP), the European Committee on Future Accelerators (ECFA), and the Asian Committee on Future Accelerators (ACFA) all issued reports endorsing the linear collider as the next major project that should be undertaken and stating that the project should be international from the start.
- In February 2001, ICFA asked the International Linear Collider Technical Review Committee (ILC-TRC) to assess the technologies for development of the ILC. This panel issued its report in 2003.
- In February 2002, ICFA established an International Linear Collider Steering Group (ILCSG) to help develop a roadmap for the ILC and to monitor and coordinate R&D activities in this area. Its responsibilities include explaining the intrinsic scientific importance of the project; defining the sci-

entific roadmap for the project; monitoring and making recommendations for the coordination of R&D efforts for the accelerator; and identifying models for international collaboration in the construction of the ILC facility. In addition, physicists in Asia, Europe, and North America formed regional ILC steering groups.

- In 2002, the Consultative Group on High-Energy Physics of the Organisation for Economic Cooperation and Development (OECD) Global Science Forum endorsed an international linear collider as the next major high-energy physics project, to be operated concurrently with the LHC.

- In the fall of 2003, ILCSG set up an International Technology Recommendation Panel (ITRP) to select a technology for the ILC.

- In January 2004, the OECD Committee for Scientific and Technology Policy issued a Ministerial Statement noting the worldwide consensus of the scientific community that an electron-positron linear collider should be the next major accelerator-based facility in particle physics.

- In March 2004, a special task force of ILCSG reported on a framework for an international organization to develop the design of the ILC. The report recommended the formation of a Global Design Effort (GDE) that would turn the selected technology for the ILC into a conceptual design and then into a design ready for construction.

- In August 2004, ITRP unanimously recommended that the ILC design incorporate the superconducting radio-frequency technology. This recommendation was immediately adopted by ILCSG and ICFA and has been accepted by the research communities of all three regions. Immediately following the selection of the technology for the ILC, the ILCSC initiated the process for the GDE.

- In March 2005, ILCSG and ICFA selected a director for the GDE. The director is coordinating activity on the project worldwide, but at present there is no centralized organization. Instead, a worldwide network with regional leaders reporting to the GDE director is being established, along with a work plan for this effort in each region.

The budgets of the science agencies in the United States, Japan, and Europe have included, directly or indirectly, R&D activities in support of the proposed ILC for a number of years. For FY2006 the U.S. budget for ILC R&D is about $25 million, with roughly similar amounts being spent in the other regions. The scientific excitement and enthusiasm for the ILC are such that all of the countries have agreed to support continuing R&D for the ILC without a commitment at this time to proceeding with construction of the project.

The GDE is working to develop credible estimates for the scope and cost of the project. The credibility of the estimates must be tested though rigorous and transparent reviews by government funding bodies, so that a high level of consensus on the expected costs can be reached. There will need to be an agreed-upon and transparent method for converting these costs according to the different accounting systems of the various national funding bodies, especially since factors such as personnel costs and contingency are treated differently in different regions.

In moving toward international bids to host the facility, it will be helpful to have certain agreed-upon elements. While regional documents differ in detail, the following checklist identifies major issues that are common concerns.

1. *Legal basis.* Due to the size and likely duration of the ILC, the international collaboration will require a durable legal structure. The legal basis for the collaboration could take the form of a treaty or an international executive agreement. It also could take a relatively novel form, such as a special-purpose, nonprofit corporation.

2. *Governance structure.* The governance structure will set out the rights of the funding nations to participate in major project decisions. Ultimately, the selection of a governance structure will be closely linked to decisions on cost sharing and site. The governance plan needs to define how voting is linked to contributions and how decisions will be reached on, for example, upgrade and/or termination plans.

3. *Project management organization.* An effective ILC organization requires strong centralized management during both construction and the subsequent operations phase. In particular, the management organization needs to exercise budgetary control and undertake construction management, ongoing site management, safety and security oversight, and personnel management.

4. *Personnel management.* While the size of the staff and other individuals on-site at any time will depend on the arrangements established for remote users, there will need to be uniform policies for the management of all personnel at the laboratory. In particular, it is important that the host can ensure equity in the treatment of all participants.

5. *Cost sharing.* The sharing of costs among the funding participants will probably be guided by a combination of a formula approach and extra premiums tied to special benefits. The governments or funding agencies of participating countries will need to agree to this structure.

6. *Financial management.* Management of finances will be of critical concern to all the participating governments. As a starting point, the ILC project organization will need to establish a set of guidelines for costing

the project and the contributions of the participants. This will ensure that all costs have been properly accounted for and that the costs (including in-kind contributions) are allocated on a fair and consistent basis. The ILC project organization will need to establish effective controls over budgeting, procurement, quality control, design changes, and contingency reserves. The ILC organization also will need to ensure that the internationally agreed-upon cash contributions are funded on a timely basis so as to avoid schedule delays and associated cost increases. Finally, the financial management procedures will need to ensure timely and transparent accountability on the financial status of the project to the national sponsors.

7. *Procurement strategy.* The rules for procurement and relationships with industry vary greatly from country to country. When components are to be built and funded in a particular region, the rules applicable for that region should be applied. Quality control remains a central management concern and must be coordinated even for nationally contributed components.

8. *Site selection criteria.* The siting of the ILC facility ultimately will be a decision made by senior governmental policy makers, but the process can be facilitated by development of a set of technical and management criteria for site selection.

9. *Experimental program management.* Plans will be needed for how the experimental program is to be managed and how remote participants will have access to the data and possibly control of the experiments from remote centers, as well as access in terms of visits to the site.

10. *Information management and dissemination.* The ILC, like the LHC, will generate huge volumes of data. Management of these data and their intellectual content will require special arrangements affecting such issues as data management, intellectual property, and information dissemination. The international nature of the laboratory will require that the data be accessible in a timely fashion to remote users as well as to those at the site.

If the United States prepares a bid to host the ILC, the issues listed above will need to be addressed for the bid to have a good chance of success. These issues are not unique to the United States but will also need to be addressed in the bids of other countries.

B

Charge to the Committee

At the dawn of the 21st century, elementary particle physics is poised to address some of the most basic questions in science. Obtaining the answers to these questions will require a global effort of great scale and complexity. The committee is charged to construct a plan for U.S. participation in this effort. In particular, the committee will

1. Identify, articulate, and prioritize the scientific questions and opportunities that define elementary particle physics.
2. Recommend a 15-year implementation plan with realistic, ordered priorities to realize these opportunities.

C

Committee Meeting Agendas

FIRST MEETING
WASHINGTON, D.C.
NOVEMBER 30-DECEMBER 1, 2004

Tuesday, November 30

Closed Session

8:15 a.m.	Introductions
	—H. Shapiro, S. Dawson
8:30	Committee composition and balance discussion
	—T. Meyer, BPA

Open Session

9:30	The coming revolutions in particle physics
	—C. Quigg, Fermilab
10:15	Break
10:30	Particle physics on the edge
	—J. Lykken, Fermilab
11:15	Strategies for discovery
	—P. Drell, SLAC
Noon	Lunch

12:30 p.m.	Purpose of and goals for the study
	—M. Turner, NSF, and R. Staffin, DOE
1:30	Setting scientific priorities
	—P. Looney, OSTP
2:00	Astronomy and astrophysics priority-setting
	—C. McKee, UC Berkeley (by telephone)
2:30	Break
3:00	Particle Physics Project Prioritization Panel (P5)
	—A. Seiden, UC Santa Cruz
3:30	Personal perspective
	—B. Barish, Caltech
4:15	Break
4:30	Public comment session
	—D. Bortoletto and M. Tuts, APS/Division of Particles and Fields (DPF) (organizers)
5:30	Adjourn

Wednesday, December 1

Closed Session

9:00 a.m.	Backdrop for this report
	—J. Bagger
9:45	International perspectives
	—P. Burrows, T. Kajita
10:30	Break
10:45	National investments in particle physics
	—J. Hezir
11:30	Discussion
12:30 p.m.	Adjourn and lunch

SECOND MEETING
MENLO PARK, CALIFORNIA
JANUARY 31-FEBRUARY 1, 2005

Monday, January 31

Closed Session

| 8:00 a.m. | Plans for this meeting |
| | —H. Shapiro, S. Dawson |

Open Session

8:30	Science accessed by the LHC
	—I. Hinchliffe, LBNL
9:00	Discussion
9:15	Break
9:30	Science reach of a linear collider and why it matters
	—J. Hewett, SLAC, and H. Murayama, Berkeley
10:45	Discussion
11:00	Break

Closed Session

11:15	Discussion

Open Session

12:30 p.m.	Lunch
1:15	Opportunities for and relevance of studying B physics
	—R. Cahn, Lawrence Berkeley National Laboratory
1:45	Discussion
2:00	Opportunities for and relevance of studying neutrinos
	—B. Kayser, Fermilab
2:30	Discussion
2:45	Break
3:00	Connections to astrophysics and cosmology
	—S. Kahn, SLAC/Kavli Institute for Particle Astrophysics and Cosmology
3:30	Discussion
3:45	Visions for the SLAC future
	—J. Dorfan, SLAC
4:15	Discussion
4:30	Break
4:45	Public comment session
	—J. Jaros and W. Carithers, APS/DPF (organizers)
5:45	Adjourn

Tuesday, February 1

Open Session

7:30 a.m.	Tour of SLAC

Closed Session

9:00   Committee discussions
12:30 p.m.   Adjourn

THIRD MEETING
BATAVIA, ILLINOIS
MAY 16-17, 2005

Monday, May 16

Closed Session

8:00 a.m.   Plans for the meeting
               —H. Shapiro, S. Dawson

Open Session

8:30   Welcome, purpose of the meeting
               —H. Shapiro, S. Dawson
8:35   The U.S. national program
               —M. Witherell, Fermilab
9:05   International cooperation and coordination in Germany
               —A. Wagner, DESY
9:35   International cooperation and coordination in Japan
               —Y. Totsuka, KEK
10:05   International cooperation and coordination in the
               United Kingdom
               —I. Halliday, Particle Physics and Astronomy Research Council
10:30   Discussion
10:45   Break
11:00   The ILC accelerator R&D program
               —S. Holmes, Fermilab
11:30   Discussion
Noon   Lunch
1:00 p.m.   Accelerator-based neutrino programs
               —G. Feldman, Harvard
1:30   Discussion

Closed Session

1:45   Committee discussions

Open Session

3:00	Cosmology and astrophysics
	—E. Kolb, University of Chicago
3:30	Discussion
4:00	Break
4:15	Visions for the Fermilab future
	—P. Oddone, Fermilab
4:45	Discussion
5:15	Public-comment session
	—M. Carena, Y.K. Kim, J. Lykken (organizers)
6:15	Adjourn

Tuesday, May 17

Open Session

7:30 a.m.	Tour of Fermilab

Closed Session

9:00	Discussion
Noon	Lunch
1:00 p.m.	Subcommittee breakout sessions
3:00	Reconvene; group discussion
4:00	Adjourn

FOURTH MEETING
ITHACA, NEW YORK
AUGUST 1-3, 2005

Monday, August 1

Closed Session

9:00 a.m.	Convene
	—H. Shapiro, S. Dawson
9:05	Subcommittee breakout sessions
5:00 p.m.	Adjourn

Tuesday, August 2

Closed Session

8:00 a.m.	Welcome and plans for the meeting —H. Shapiro, S. Dawson
8:15	Discussion
9:30	Break

Open Session

9:45	Welcome, purpose of the meeting —H. Shapiro, S. Dawson
10:00	TeV-scale physics —N. Arkani-Hamed, Harvard
10:30	HEPAP subpanel on LHC/ILC synergy —J. Lykken, Fermilab
11:00	Discussion
11:30	The ILC Global Design Effort —B. Barish, Caltech
Noon	Discussion
12:15 p.m.	Lunch
1:15	Perspectives from CERN —R. Aymar, CERN
1:45	Discussion
2:00	Role of the International Committee on Future Accelerators —J. Dorfan, SLAC
2:30	Discussion
2:45	Break

Closed Session

3:00	Discussion

Open Session

4:30	Visions for the Cornell future —M. Tigner, Cornell
5:00	Discussion
5:15	Public-comment session —R. Polling and I. Shipsey (organizers)
6:15	Adjourn

Wednesday, August 3

Open Session

7:30 a.m.   Tour of CLEO, CHESS, and other facilities

Closed Session

9:00   Discussion
Noon   Lunch
1:00 p.m.   Discussion
4:00   Adjourn

**FIFTH MEETING
WASHINGTON, D.C.
DECEMBER 6, 2005**

Tuesday, December 6

Open Session

8:00 a.m.   Welcome
              —H. Shapiro, S. Dawson
8:05   Accelerator-based neutrino experiments: Fermilab
              —G. Feldman, Harvard
8:20   Accelerator-based neutrino experiments: J-PARC
              —T. Kajita, University of Tokyo
8:40   Discussion

Closed Session

9:00   Discussion
10:30   Break
11:00   Discussion
Noon   Lunch
1:00 p.m.   Discussion
6:00   Adjourn

SIXTH MEETING
WASHINGTON, D.C.
JANUARY 23, 2006

Monday, January 23

Open Session

8:00   Welcome; plans for the meeting
            —H. Shapiro, S. Dawson
8:10   Perspectives from Fermilab: An update
            —P. Oddone, Fermilab
8:30   Discussion

Closed Session

9:30   Discussion
Noon   Lunch
1:00 p.m.   Discussion
6:00   Adjourn

D

Biographical Sketches of Committee Members and Staff

COMMITTEE MEMBERS

Harold T. Shapiro, *Chair*

Dr. Shapiro is president emeritus of Princeton University and a professor of economics and public affairs in the Woodrow Wilson School. He received his Ph.D. in economics from Princeton in 1964 and his bachelor's from McGill University in 1956. He served as president of the University of Michigan from 1980 to 1988. Dr. Shapiro's expertise is in econometrics. A member of the Institute of Medicine, he has been widely recognized and decorated for his shrewd judgment in policy situations, from his chairing of the National Bioethics Advisory Committee under President Clinton to his service on the President's Council of Advisors on Science and Technology under President Bush. Other distinctions include chairing the Association of American Universities, service on the board of directors of the National Bureau of Economic Research, Inc., and the board of trustees of the Universities Research Association, Inc. He has chaired and served on numerous NRC committees, including the most recent Committee on the Organizational Structure of the National Institutes of Health. Dr. Shapiro was recently awarded the 2006 William D. Carey Lecture of the American Association for the Advancement of Science for his leadership in science policy.

Sally L. Dawson, *Vice-chair*

Dr. Dawson is chair of the Physics Department at Brookhaven National Laboratory and an adjunct professor at the Institute for Theoretical Physics at SUNY at Stony Brook. She received her Ph.D. from Harvard University in 1981 under Howard M. Georgi. Dr. Dawson was recently chair of the American Physical Society's Division of Particle and Fields, the primary professional society for elementary particle physics, a position to which she was elected by her peers. Her primary scientific expertise is in the area of theoretical high-energy physics, specializing in studies of the Higgs boson, electroweak symmetry breaking, and physics beyond the Standard Model. She is a fellow of the American Physical Society and was awarded the Woman of the Year in Science by the Town of Brookhaven in 1995. Her committee service includes the High Energy Physics Advisory Panel of the Department of Energy and the National Science Foundation, the American Physical Society Committee on the Status of Women in Physics, and the International Committee on the Future of Accelerators.

Norman R. Augustine

Mr. Augustine retired in 1997 as chair and chief executive officer of Lockheed Martin Corporation. Previously he served as chair and CEO of the Martin Marietta Corporation. Upon retiring he joined the faculty of the Department of Mechanical and Aerospace Engineering at Princeton University. Earlier in his career he had served as under secretary of the Army and before that as assistant director of Defense Research and Engineering. Mr. Augustine served 9 years as chairman of the American Red Cross. He has also been president of the American Institute of Aeronautics and Astronautics and served as chairman of the Henry M. "Scoop" Jackson Foundation for Military Medicine. He has served as a trustee of the Massachusetts Institute of Technology, Johns Hopkins University, and Princeton University. He serves on the President's Council of Advisors on Science and Technology and is a former chairman of the Defense Science Board. He currently serves on the corporate boards of Black and Decker, Procter and Gamble, and ConocoPhillips. He has been presented the National Medal of Technology and the Department of Defense's highest civilian award, the Distinguished Service Medal, five times; in 2006, he was awarded the National Academy of Sciences' Public Welfare Medal. Mr. Augustine holds an M.S. in aeronautical engineering from Princeton University. He has been elected to membership in the National Academy of Sciences and the National Academy of Engineering; he served as chairman of the National Academy of Engineering for 2 years.

Jonathan A. Bagger

Dr. Bagger is Krieger-Eisenhower Professor in the Department of Physics and Astronomy at the Johns Hopkins University. He received his Ph.D. from Princeton University in 1983. His primary research interests are in theoretical particle physics, particularly in the theory and phenomenology of supersymmetry, supergravity, and superstrings. Dr. Bagger has twice been a member of the Institute for Advanced Study in Princeton, New Jersey. He held a Sloan Foundation Fellowship and a National Science Foundation Presidential Young Investigator award. He is vice-chair of the Department of Energy/National Science Foundation High Energy Physics Advisory Panel and a member of the Fermi National Accelerator Laboratory board of overseers. He has served on the Stanford Linear Accelerator Center Scientific Policy Committee and as chair of the Division of Particles and Fields of the American Physical Society. Dr. Bagger is a member of the Board on Physics and Astronomy. He also helped organize the first "Frontiers of Science" symposium of the National Academies.

Philip N. Burrows

Dr. Burrows is the professor of accelerator physics at the John Adams Institute, University of Oxford. He received his Ph.D. in particle physics in 1988 from Oxford University. His areas of expertise include experimental particle physics and accelerator science and technology. He is one of the world's experts on the science and technology possibilities for future accelerator-based particle physics projects. He has been involved in the design and testing of several fast-feedback systems that are critical for future accelerator projects. He also co-chaired a working group on quantum chromodynamics at the Snowmass 2001 meeting of the high-energy physics community. As a member of the European particle physics community who has also participated in experiments at Stanford Linear Accelerator Center, Professor Burrows is familiar with the international context of particle physics.

Sandra M. Faber

Dr. Faber is a professor of astronomy at the University of California at Santa Cruz with the Lick Observatory. Her research focuses on the formation and evolution of galaxies and the evolution of structure in the universe. She utilizes ground-based optical data obtained with the Lick 3-meter and Keck 10-meter telescopes. She is a member of the Wide-Field Camera (I) Team of the Hubble Space Telescope. Dr. Faber is also a core member of the Deep Extragalactic Evolutionary Probe project, a large-scale survey of distant, faint field galaxies using the Keck twin telescopes and the Hubble Space Telescope. She is an elected member of the National Academy of Sciences and has served as a member of the National Research Council Astronomy Survey Study, the Board on Physics and Astronomy, the Committee on Astronomy and Astrophysics, and the Committee on Physics of the Universe. Dr. Faber is also a current member of Fermi National Accelerator Laboratory's board of overseers.

Stuart J. Freedman

Dr. Freedman is a Luis W. Alvarez Chair of Experimental Physics at the University of California at Berkeley with a joint appointment to the Nuclear Science Division of the Lawrence Berkeley National Laboratory. He received his Ph.D. from the University of California at Berkeley in 1972. His research experience spans nuclear and atomic physics, neutrino physics, and small-scale experiments in particle physics, all focused on fundamental questions about the Standard Model. He recently co-chaired the American Physical Society's physics of neutrinos study and currently co-chairs the National Research Council's Rare Isotope Science Assessment Committee. Dr. Freedman is a member of the National Academy of Sciences.

Jerome I. Friedman

Dr. Friedman is an institute professor emeritus at the Massachusetts Institute of Technology. He received his Ph.D. in physics from the University of Chicago in 1956. He completed his post-doctoral work at the High Energy Physics Laboratory of Stanford University before joining the MIT faculty. His work at the Stanford Linear Accelerator Center was famously celebrated with the joint award of the 1990 Nobel prize in physics for demonstrating the substructure of the proton, a discovery that helped confirm the quark model of hadrons. Dr. Friedman is an expert in experimental particle physics and has served as head of the Physics Department and director of the Laboratory of Nuclear Science at MIT and on many advisory panels, including the joint Department of Energy/National Science Foundation High Energy Physics Advisory Panel, the Board of the University Research Association, the Board on Physics and Astronomy of the National Research Council, and as chair of the Scientific Policy Committee of the Superconducting Collider. He is a member of the National Academy of Sciences.

David J. Gross

Dr. Gross is director of the Kavli Institute for Theoretical Physics at the University of California at Santa Barbara. He received his Ph.D. from the University of California at Berkeley in 1966. Dr. Gross was co-discoverer of the asymptotic freedom of non-Abelian gauge theories and played a central role in initiating quantum chromodynamics as the modern theory of strong interactions. His incisive papers on field theory and particle physics have been widely influential. Recently, he has made seminal contributions to the theory of superstrings. He is a fellow of the American Physical Society and the American Association for the Advancement of Science and a member of National Academy of Sciences. He is the recipient of the J.J. Sakurai Prize of the American Physical Society, a MacArthur Foundation fellowship prize, the Dirac medal of the International Center for Theoretical Physics, the Oskar Klein medal, and the Harvey prize of the Israel Institute of Technology. In 2004 David Gross was selected to receive France's highest scientific honor, the Grande Médaille d'Or, for his contributions to the understanding of fundamental physical reality. Dr. Gross was awarded the 2004 Nobel prize in physics for the discovery of asymptotic freedom in the theory of strong interactions.

Joseph S. Hezir

Mr. Hezir is the cofounder and managing partner of the EOP Group, Inc., a consulting firm that specializes in federal government regulatory strategy development and budget policy. He previously served 18 years in the White House Office of Management in positions of increasing responsibility, serving for 6 years as deputy associate director for energy and science. He has also served on a number of advisory bodies, including the National Aeronautics and Space Administration's Advisory Council and the Metropolitan Area Board of Directors for the Red Cross. He also was a member of the National Research Council's Burning Plasma Assessment Committee.

Norbert Holtkamp

Dr. Holtkamp is director of the Accelerator Systems Division for the Spallation Neutron Source at Oak Ridge National Laboratory. The facility is under construction and will be completed in 2006. He received his Ph.D. from the Technical University at Darmstadt. His research interests include high-energy colliders, linear accelerators, storage rings, and accelerator-based neutrino physics. Dr. Holtkamp was a senior staff member at the Deutsches Elektronen-Synchrotron Laboratory in Germany and was also a member of the technical staff at Fermi National Accelerator Laboratory. He has served on panels of the joint Department of Energy/National Science Foundation High Energy Physics Advisory Panel and has chaired technical advisory studies examining the feasibility of various large projects.

Takaaki Kajita

Dr. Kajita is a professor of physics and director of the Research Center for Cosmic Neutrinos at the Institute for Cosmic Ray Research at the University of Tokyo. He is an expert on neutrino physics and proton decay and was a leader of the Super Kamiokande experiment that first observed evidence of neutrino oscillations. He is the author of many articles on the topic of non-accelerator-based particle physics, including several aimed at the broader public. He is also an organizer of the American Physical Society's physics of neutrinos study and a member of the Particle and Nuclear Astrophysics and Gravitational International Committee of the International Union of Pure and Applied Physics, which discusses the international organization and coordination of particle physics. He has participated in many Japanese and U.S.-based studies on the future of particle physics.

Neal F. Lane

Dr. Lane is the Malcolm Gillis University Professor, a professor in the Department of Physics and Astronomy, and the senior fellow for science and technology at the James A. Baker III Institute for Public Policy at Rice University. He earned his Ph.D. in physics from the University of Oklahoma in 1964. His research expertise is in the area of atomic and molecular physics. Dr. Lane has served as provost of Rice University, chancellor of the University of Colorado at Colorado Springs, and director of the Division of Physics at the National Science Foundation. He also directed the National Science Foundation from 1993 to 1998 and served as a member (ex officio) of the National Science Board. From 1998 to 2001, he served as assistant to the President for science and technology and director of the White House Office of Science and Technology Policy. Dr. Lane is currently serving on the National Academies' Policy and Global Affairs Committee and has been a member of multiple past panels covering atomic and molecular physics. He has a distinguished academic and teaching career in physics in addition to his years of outstanding administrative service in national science policy.

Nigel S. Lockyer

Dr. Lockyer is a professor of physics at the University of Pennsylvania. He received his Ph.D. from the Ohio State University in 1984. Dr. Lockyer is an experimental high-energy physicist with experience that spans both electron-positron and proton-antiproton colliders. Dr. Lockyer has broad experience in large-scale high-energy physics collaborations as well as small-scale experiments. He is the former co-spokesperson of the CDF experiment at Fermi National Accelerator Laboratory and was spokesperson for the Mark II experiment at Stanford Linear Accelerator Center. His research has focused on measurements of bottom quark properties.

Sidney R. Nagel

Dr. Nagel is the Stein-Freiler Distinguished Service Professor in the Department of Physics at the University of Chicago. He received his Ph.D. in physics from Princeton in 1974. Dr. Nagel served as director of the University of Chicago Materials Research Laboratory from 1987 until 1991. His research expertise lies in the area of nonlinear and disordered systems far from equilibrium, including jamming, structural glasses, granular materials, and fluids. He is a member of the National Academy of Sciences, a fellow of the American Physical Society, the American Association for the Advancement of Science, and the American Academy of Arts and Sciences and was the recipient of the 1999 Oliver E. Buckley Condensed Matter Physics Prize. He has also served as chair of the American Physical Society's Division of Condensed Matter Physics.

Homer A. Neal

Dr. Neal is the Samuel A. Goudsmit Distinguished University Professor of Physics at the University of Michigan. He received his Ph.D. from the University of Michigan in 1966. His current research is based on the D0 experiment at Fermi National Accelerator Laboratory and the ATLAS experiment at the Large Hadron Collider in Geneva. His expertise is in detector development, software development, spin physics, top quark studies, and inclusive hadron physics. Dr. Neal is the institutional leader of the University of Michigan team for ATLAS at the LHC. He has held many administrative posts at the University of Michigan, including interim president (1996-1997) and vice president for research (1993-1996). He has served as a regent of the Smithsonian Institution, as a member of the National Science Board, and as chairman of the National Science Foundation's Physics Advisory Committee. He has served on the Boards of Argonne National Laboratory, Fermi National Acceleratory Laboratory, the SSC Laboratory, and Oak Ridge National Laboratory.

J. Ritchie Patterson

Dr. Patterson is a professor of physics at Cornell University. She received her Ph.D. in particle physics from the University of Chicago in 1990. Dr. Patterson has been involved with the CLEO electron-positron experiment studying b- and c-quark physics, and her research has covered many aspects of accelerator physics, especially simulations of beam dynamics. Dr. Patterson has been a member of the joint

Department of Energy/National Science Foundation High Energy Physics Advisory Panel, as well as the Particle Physics Project Prioritization Panel, the Fermi National Accelerator Laboratory's Long-Range Planning Committee, the International Organizing Committee of the Worldwide Study of Physics and Detectors for Future Linear e^+e^- Colliders, and the Department of Energy/National Science Foundation High Energy Physics Facilities Committee. She is also a member of the CMS experiment at the Large Hadron Collider in Geneva. She is a leading expert on the technical and scientific issues of the proposed International Linear Collider project.

Helen R. Quinn

Dr. Quinn is Education and Public Outreach Manager and a senior staff scientist at the Stanford Linear Accelerator Center. She received her Ph.D. from Stanford University in 1967 in elementary particle physics. Dr. Quinn's accomplishments include authoring the first paper to discuss the unification of coupling constants in a grand unified theory, investigating ground-breaking phenomenological analysis of CP violation in B meson systems, and introducing the principle of quark-hadron duality. She is also well known for her work on science education standards in the state of California. Dr. Quinn is a member of the National Academy of Sciences and received the 2002 Dirac Medal for her seminal contributions to the field. She is also president of the Contemporary Physics Education Project, a non-profit group that produces materials discussing modern physics for high school and college use.

Charles V. Shank

Dr. Shank has served as Director of Ernest Orlando Lawrence Berkeley National Laboratory in Berkeley, California, since September 1989 until his recent retirement. He received his Ph.D. from the University of California at Berkeley in 1969. In addition to his duties as Laboratory Director, Dr. Shank has a unique triple appointment as professor at the University of California at Berkeley in the Department of Physics, Department of Chemistry, and Department of Electrical Engineering and Computer Sciences. Dr. Shank's scientific and service contributions in optical science and engineering have been recognized through honors that include the R.W. Wood Prize of the Optical Society of America, the David Sarnoff and Morris E. Leeds awards of the Institute of Electrical and Electronics Engineers, and the Edgerton Award of the International Society for Optical Engineering. He was the chair of the National Research Council's Committee on Optical Science and Engineering, which published its report in 1998.

Paul J. Steinhardt

Dr. Steinhardt is the Albert Einstein Professor of Science in the Department of Physics at Princeton University. He has made outstanding contributions in cosmology and condensed matter physics. He is a leading expert on inflationary cosmology and other events in the very early universe. His work led to the first inflationary models for the universe, to the discovery that quantum fluctuations could seed galaxy formation, and to new observational tests of these models. Using concepts of string theory, he has developed an alternative, known as the cyclic model of the universe. He also introduced the concept of quasicrystals and pioneered the study of their structural and elastic properties in condensed-matter physics. He is a fellow of the American Physics Society and an elected member of the National Academy of Sciences.

Harold E. Varmus

Dr. Varmus is president of the Memorial Sloan-Kettering Cancer Center in New York and a former director of the National Institutes of Health. He received an M.D. from Columbia University and an M.A. in English from Harvard University. He served for many years on the faculty of the University of California at San Francisco before directing the NIH under President Clinton for 6 years. Dr. Varmus is co-recipient of a Nobel prize in physiology or medicine for studies of the genetic basis of cancer and is a member of the National Academy of Sciences and the Institute of Medicine. He received the National Medal of Science in 2001.

Edward Witten

Dr. Witten is a professor of physics in the School of Natural Sciences at the Institute for Advanced Study in Princeton. He earned his Ph.D. in 1976 from Princeton University in high-energy physics. Dr. Witten is a theoretical physicist whose solution of outstanding problems in string theory greatly advanced its status as one of the leading candidates for the grand unified theory of elementary particle physics. He has received the Dirac medal of the International Center for Theoretical Physics, the Fields medal of the International Mathematical Union, the Madison medal of Princeton University, and the Einstein medal of the Albert Einstein Society. He has been elected a member of the National Academy of Sciences.

NRC STAFF

Donald C. Shapero, Director, Board on Physics and Astronomy

Dr. Shapero received a B.S. degree from the Massachusetts Institute of Technology (MIT) in 1964 and a Ph.D. from MIT in 1970. His thesis addressed the asymptotic behavior of relativistic quantum field theories. After receiving the Ph.D., he became a Thomas J. Watson Postdoctoral Fellow at IBM. He subsequently became an assistant professor at American University, later moving to Catholic University, and then joining the staff of the National Research Council in 1975. Dr. Shapero took a leave of absence from the NRC in 1978 to serve as the first executive director of the Energy Research Advisory Board at the Department of Energy. He returned to the NRC in 1979 to serve as special assistant to the president of the National Academy of Sciences. In 1982, he started the NRC's Board on Physics and Astronomy (BPA). As BPA director, he has played a key role in many NRC studies, including the two most recent surveys of physics and the two most recent surveys of astronomy and astrophysics. He is a member of the American Physical Society, the American Astronomical Society, and the International Astronomical Union. He has published research articles in refereed journals in high-energy physics, condensed-matter physics, and environmental science.

Timothy I. Meyer, Senior Program Officer, Board on Physics and Astronomy

Dr. Meyer is a senior program officer at the NRC's Board on Physics and Astronomy. He received a Notable Achievement Award from the NRC's Division on Engineering and Physical Sciences in 2003 and a Distinguished Service Award from the National Academies in 2004. Dr. Meyer joined the NRC staff in 2002 after earning his Ph.D. in experimental particle physics from Stanford University. His doctoral thesis concerned the time evolution of the B meson in the BaBar experiment at the Stanford Linear Accelerator Center. His work also focused on radiation monitoring and protection of silicon-based particle detectors. During his time at Stanford, Dr. Meyer received both the Paul Kirkpatrick and the Centennial Teaching awards for his work as an instructor of undergraduates. He is a member of the American Physical Society, the American Association for the Advancement of Science, the Materials Research Society, and Phi Beta Kappa.

David B. Lang, Research Associate, Board on Physics and Astronomy

Mr. Lang is a research associate at the NRC's Board on Physics and Astronomy (BPA). He received a B.S. in astronomy and astrophysics from the University of Michigan in 2002. His senior thesis concerned surveying very young galaxies in a field beside the irregular galaxy Sextans-A using the Hubble Space Telescope. His mentors were Robbie Dohm-Palmer, University of Minnesota, and Mario Mateo, University of Michigan. Mr. Lang came to the BPA after having worked in an intellectual property law firm in Arlington, Virginia, for 2 years and began at the BPA as a research assistant. He performs supporting research for studies ranging from radio astronomy to materials science and recently received the "Rookie" award of the NRC's Division on Engineering and Physical Sciences. He is a member of the American Astronomical Society.